Thomas Harriot: Science and Discovery in the English Renaissance

This volume sheds new light on one of the most remarkable polymaths of the English Renaissance. It offers original perspectives not only on Harriot's personal achievements in mathematics and natural philosophy but also on the wider realms of exploration, colonial ambition, and philosophical debate in which he earned the attention and respect of contemporaries in and far beyond the socially elevated circles of his two great patrons, first Walter Ralegh and then Henry Percy, the ninth Earl of Northumberland.

Harriot's sixteenth-century world was one of unprecedented expansion in both scientific understanding and the discovery of new lands and peoples. The essays gathered here bring out forcefully the effect of this expanding vision, encapsulated in Harriot's *Briefe and true report of the new found land of Virginia* (1588), the first detailed description of America to be published in the English language. In addition to an essay by a recent biographer of Harriot, the volume contains reworked versions of seven Thomas Harriot Lectures, an annual lecture series inaugurated in 1990 in Oriel College, Oxford. It follows two earlier volumes of Harriot Lectures, also edited by Robert Fox, that appeared in 2000 and 2012.

Robert Fox is Emeritus Professor of the History of Science at the University of Oxford and an honorary fellow of Oriel College. His main research interests are in European science, technology, and medicine since the eighteenth century. His recent books include *The Savant and the State. Science and Cultural Politics in Nineteenth-Century France* (2012) and *Science without Frontiers. Cosmopolitanism and National Interests in the World of Learning, 1870–1940* (2016).

Thomas Harriot: Science and Discovery in the English Renaissance

Edited by Robert Fox

Routledge
Taylor & Francis Group

LONDON AND NEW YORK

First published 2023
by Routledge
4 Park Square, Milton Park, Abingdon, Oxon OX14 4RN

and by Routledge
605 Third Avenue, New York, NY 10158

Routledge is an imprint of the Taylor & Francis Group, an informa business

British Library Cataloguing-in-Publication Data
A catalogue record for this book is available from the British Library

Library of Congress Cataloging-in-Publication Data
Names: Fox, Robert, 1938- editor.
Title: Thomas Harriot: science and discovery in the English Renaissance / edited by Robert Fox. Other titles: Science and discovery in the English Renaissance
Description: Abingdon, Oxon ; New York, NY : Routledge, 2023. | Series: Thomas Harriot Lectures |
Includes bibliographical references and index.
Identifiers: LCCN 2022030368 (print) | LCCN 2022030369 (ebook) | ISBN 9780367561376 (hardback) | ISBN 9780367561369 (paperback) | ISBN 9781003096580 (ebook)
Subjects: LCSH: Harriot, Thomas, 1560-1621. |
Explorers--Great Britain–Biography. | Virginia--Discovery and exploration--English. | Great Britain--History--Elizabeth, 1558-1603--Biography. | Scientists--Great Britain--Biography. | Mathematics--England--History--16th century.
Classification: LCC Q143.H36 .T475 2000 (print) |
LCC Q143.H36 (ebook) | DDC 509.2 [B]--dc23/eng/20220719
LC record available at https://lccn.loc.gov/2022030368
LC ebook record available at https://lccn.loc.gov/2022030369

ISBN: 978-0-367-56137-6 (hbk)
ISBN: 978-0-367-56136-9 (pbk)
ISBN: 978-1-003-09658-0 (ebk)

DOI: 10.4324/9781003096580

Typeset in Times New Roman
by MPS Limited, Dehradun

Dedicated to the memory of Jacqueline Stedall (1950-2014)

Contents

Figures

pouch, nuts, fish, corn, and shells; but also landscape, sky, and clouds. In the hand-colored copy of this German print the woman's body is white; her hair is blonde, her eyes blue, and with red lipstick—very unlikely colors to be encountered among American Indians. From Thomas Harriot, *Wunderbarliche, doch warhafftige Erklärung vonder Gelegenheit und Sitten der Wilden in Virginia* (1590), North Carolina Collection, University of North Carolina Chapel Hill FVCC970.1 H28w. Courtesy of Wilson Library Special Collections, University of North Carolina   195

Contributors

Polly Allingham is a journalist and writer.

Robyn Arianrhod is a science writer, and a mathematician affiliated with Monash University's School of Mathematics (her field of research is general relativity). Her articles and reviews have appeared in *Cosmos, the Conversation, Times Higher Education, Washington Post,* and elsewhere, and her books include *Einstein's Heroes: Imagining the World through the Language of Mathematics; Seduced by Logic: Émilie du Châtelet, Mary Somerville and the 'Newtonian Revolution';* and, most recently, *Thomas Harriot: A Life in Science.*

Philip Beeley is Research Fellow in the Faculty of History and Fellow and Tutor of Linacre College, Oxford. A former President of the British Society for the History of Mathematics, he specializes in the history of science and mathematics in early modern Europe with a particular focus on John Wallis and Gottfried Wilhelm Leibniz. Not least through the work of these two figures he has always had a strong research interest in the scientific heritage of Thomas Harriot.

Daniel Carey is Director of the Moore Institute at the University of Galway. His publications on travel, exploration, and colonialism include, as editor, *Richard Hakluyt and Travel Writing in Early Modern Europe* (2012); *The Postcolonial Enlightenment: Eighteenth-Century Colonialism and Postcolonial Theory* (2009); and *Asian Travel in the Renaissance* (2004). He is General Editor, with Claire Jowitt, of Richard Hakluyt's *Principal Navigations, Voyages, Traffiques, and Discoveries of the English Nation,* in preparation for OUP in 14 volumes.

Stephen Clucas is Reader in Early Modern Intellectual History at Birkbeck, University of London. He is the Chairman of the Thomas Harriot Seminar, and from 2007 to 2019 was editor (with Stephen Gaukroger) of *Intellectual History Review.* A collection of his essays, *Memory, Magic and Natural Philosophy in the Sixteenth and Seventeenth Centuries* was published in 2011. He is currently editing (with Timothy J. Raylor)

Thomas Hobbes's *De corpore* for the Clarendon Edition of the works of Thomas Hobbes.

Felipe Fernández-Armesto is Vice-president of the Hakluyt Society and occupies the William P. Reynolds Chair for Mission in Arts and Letters at the University of Notre Dame, where he is a professor in the Departments of History and Classics. Recent books include *Out of Our Minds: What We Think and How We Think It* (2019), *A Foot in the River* (2015), and, as editor, *The Oxford Illustrated History of the World* (2019).

Robert Fox is Emeritus Professor of the History of Science at the University of Oxford and an honorary fellow of Oriel College, where since 1990 he has organized the College's annual Thomas Harriot Lecture. He has edited two earlier volumes of Harriot lectures, published by Ashgate in 2000 and 2012.

Mark Horton is Emeritus Professor at the University of Bristol, and Professor of Archaeology and Cultural Heritage at the Royal Agricultural University. His PhD (Cambridge) was on the coastal trading settlements of East Africa, trade networks, and colonizations in the Indian Ocean. He has since led excavations on pioneer European settlements including the Scottish Colony in Darien (1698–1700), the English in Bermuda (1610), and St Lucia and St Kitts (1624). Since 2009, he has worked with the Croatan Archaeological Society on the history and archaeology of the Outer Banks. He has also presented BBC Coast series (2005–16).

David Harris Sacks is Richard F. Scholz Professor of History and Humanities, Emeritus in Reed College, Portland, Oregon, USA. He has published on a variety of subjects in Renaissance and Early Modern History, focused especially on the development of political, religious and scientific discourses and practices in the Atlantic world. His researches began with the history of the city of Bristol, include work on More's *Utopia*, and have concentrated most recently on the careers of Richard Hakluyt and Thomas Harriot.

Larry E. Tise. As Wilbur and Orville Wright Distinguished Professor at the Thomas Harriot College, East Carolina University, Tise launched an ongoing programme of Harriot research, lectures, and commemorative events. His inventory of hand-coloured versions of Harriot's *Briefe and True Report of the New Found Land of Virginia* (1590) resulted in an elephantine art catalogue, *Theodor de Bry—America: The Complete Prints, 1590–1602* (2019). His other books focus on slavery and race, Benjamin Franklin, the Wright brothers, and North Carolina.

Preface

When David Quinn gave the first Thomas Harriot Lecture in Oriel College in May 1990, there was little expectation that the lecture would remain an annual event in the college's calendar more than 30 years later. In that time, two volumes of lectures have been published: *Thomas Harriot: An Elizabethan Man of Science* (2000) and *Thomas Harriot and His World: Mathematics, Exploration, and Natural Philosophy in Early Modern England* (2012), both under the imprint of Ashgate. Now that a happy collaboration with Ashgate has given way to a new association with Routledge, it is a pleasure to record the warm welcome that the current volume has received from Routledge's editorial team, in particular from Michael Greenwood and Louis Nicholson-Pallett, and from the MPS Limited production unit in India, where Radhika Bhartari, as Project Manager for the book, has handled a complex process with exemplary efficiency and courtesy.

Over the years, the lectures have reflected the evolving profile of research on Harriot and the many worlds in which he achieved distinction, and they continue to do so. The present volume, for example, offers rather more than the earlier ones on the *Briefe and true report of the new found land of Virginia* and the wider context of English and Spanish colonial aspirations within which the book was written. Despite the modest size of the original text of 1588 and the illustrated edition that Theodor de Bry published two years later, the *Report* remains a rich quarry for scholarly enquiry into the early exploration of the New World and the activities it stimulated, ranging from the practical arts of navigation and gunnery to the curiosity-led pursuits of geography, botany, and even ethnography. But Harriot's mathematical and scientific work and his place in the European mathematical tradition remain as prominent a focus as ever. Here, interest can only be expected to grow as the several thousand pages of manuscripts become more readily available through the ECHO (European Cultural Heritage Online) digitization project hosted at the Max Planck Institute for the History of Science in Berlin and led by an international team headed by two past Harriot lecturers, Matthias Schemmel in Berlin and Robert Goulding at the University of Notre Dame.

From the start, Oriel has valued its association with this wider world of Harriot scholarship. It has provided a supportive and natural setting for the lectures, which commemorate Harriot's association with St Mary Hall. It was at St Mary Hall, now part of Oriel, that he graduated in 1580, and a fine modern copy of what is commonly thought to be his portrait hangs in the Oriel dining hall. Within the college, the support of unfailingly enthusiastic Provosts has been essential, and it has been a privilege to work, successively, with the late Sir Zelman Cowan, the late Revd Professor Ernest Nicholson, Sir Derek Morris, Moira Wallace, and Neil Mendoza, Lord Mendoza. For many years, too, Max, Lord Egremont has been a wise and friendly presence. He has been generous not only in his support and the interest he has shown in the lectures in Oriel but also in his commitment to the promotion of Harriot studies on a far broader canvas. 'Harrioteers' everywhere are in his debt.

In a venture with such distant roots, losses have been inevitable, and we look back fondly and gratefully to some of the earlier lecturers who are no longer with us. In dedicating this volume to the memory of Jacqueline Stedall, we pay a special tribute to an inspirational leader among Harriot scholars in and far beyond Oxford, whose premature death we mourned during the present cycle of lectures.

Robert Fox
Oxford
August 2022

Thomas Harriot Lectures 1990–2021

The following Thomas Harriot Lectures have been delivered at Oriel College since their inauguration in 1990.

David B. Quinn, 'Thomas Harriot and the problem of America', delivered on 7 May 1990.

Gordon R. Batho, 'Thomas Harriot and the Northumberland household', delivered on 20 May 1991.

Hugh Trevor-Roper, 'Harriot's physician: Theodor de Mayerne', delivered on 14 May 1992.

Hilary Gatti, 'The natural philosophy of Thomas Harriot', delivered on 20 May 1993.

Stephen Clucas, 'Thomas Harriot and the field of knowledge in the English Renaissance', delivered on 19 May 1994.

J.A. Bennett, 'Instruments, mathematics, and natural knowledge: Thomas Harriot's place on the map of learning', delivered on 18 May 1995.

Muriel Seltman, 'Harriot's algebra: reputation and reality', delivered on 2 May 1996.

John D. North, 'Stars and atoms', delivered on 22 May 1997.

John J. Roche, 'Harriot, Oxford, and twentieth-century historiography', delivered on 14 May 1998.

Scott Mandelbrote, 'The religion of Thomas Harriot', delivered on 20 May 1999.

Jon V. Pepper, 'Thomas Harriot and the great mathematical tradition', delivered on 11 May 2000.

Robert Goulding, 'Thomas Harriot's optical researches', delivered on 24 April 2001.

Jacqueline Stedall, *'The greate invention of algebra:* Thomas Harriot's treatise on equations', delivered on 16 May 2002.

Ian Maclean, 'Thomas Harriot on combinations', delivered on 22 May 2003.

Matthias Schemmel, 'The English Galileo: Thomas Harriot and the force of shared knowledge in early modern mechanics', delivered on 10 June 2004.

John Henry, 'Why Thomas Harriot was not the English Galileo', delivered on 26 May 2005.

Stephen Pumfrey, 'Patronage, protection, and publication of scientists in the Renaissance: the strange case of Thomas Harriot', delivered on 18 May 2006.

Stephen Johnston, 'Thomas Harriot and the English experience of navigation', delivered on 17 May 2007.

Mark Nicholls, 'Last act? 1618 and the shaping of Sir Walter Raleigh's reputation', delivered on 22 May 2008.

Pascal Brioist, 'Thomas Harriot and the worlds of practice: learning from seamen and soldiers', delivered on 4 June 2009.

Surekha Davies, 'Thomas Harriot, John White, and the invention of the Algonquian Indian, 1585–1650', delivered on 19 May 2011.

Lesley Cormack, '"The whole earth, a present for a Prince": Molyneux's English globes and the creation of a global vision in Harriot's time', delivered on 31 May 2012.

David Reed Sacks, 'The true and certain discovery of the world: Thomas Harriot and Richard Hakluyt', delivered on 29 May 2014.

Stephen Clucas, 'Thomas Harriot in the twenty-first century: 25 years of the Harriot Lecture', delivered on 28 May 2015.

Philip Beeley, 'Our learned countryman. Thomas Harriot and the emergence of mathematical community in seventeenth-century England', delivered on 1 June 2016.

Mark Horton, 'Thomas Harriot, the world's first ethnographer?', delivered on 24 May 2017.

Daniel Carey, 'Harriot, Hakluyt, and Newfound Virginia', delivered on 22 May 2018.

Felipe Fernández-Armesto, "Both to love and fear us". How to found an empire in Harriot's day, delivered on 23 May 2019.

Larry E. Tise, 'Thomas Harriot's magnificent book. Creating Europe's first illustrated exploration narrative', originally planned as the 2020 lecture, delivered on 11 November 2021.

Revised versions of the lectures delivered between 1990 and 1999 were published in Robert Fox (ed.), *Thomas Harriot. An Elizabethan man of science* (Aldershot: Ashgate, 2000).

Revised versions of the lectures delivered between 2000 and 2009, with the exception of Stephen Johnston's lecture of 2007, appeared in Robert Fox (ed.), *Thomas Harriot and his world. Mathematics, exploration, and natural philosophy in early modern England* (Farnham: Ashgate, 2012).

The lectures delivered since 2014 have been reworked for publication in the present volume. The lectures by Stephen Johnston, Surekha Davies, and Lesley Cormack remain unpublished.

Introduction: Thomas Harriot. Science, Mathematics, Exploration in the English Renaissance

Robert Fox

Over three decades, Oriel College's Thomas Harriot Lectures have reflected the evolving profile of a body of Harriot scholarship that has grown spectacularly since his achievements began to attract new interest in the 1950s. The emphasis in the earlier lectures, most of them revised and collected in volumes published in 2000 and 2012, tended to be on Harriot as a mathematician, man of science, and philosopher.[1] Never far away, however, was a seemingly different Harriot, the Harriot whose skills in navigational techniques led to his involvement in Sir Walter Ralegh's colonial and commercial ventures in the New World and took him to the coast and barrier islands of what is now North Carolina, known then as Virginia, in 1585–86. In fact, it was on this Harriot, engaged in Ralegh's service since 1583, that the distinguished historian of the exploration and settlement of America, David Beers Quinn, gave the very first Thomas Harriot Lecture in 1990.[2]

A formidable challenge for Quinn and other pioneers of Harriot studies was the patchy quality of the sources with which they had to work. The published corpus of primary materials was thin and unsound. In mathematics, it amounted to no more than the *Artis analyticae praxis*, a work on the theory of equations that appeared posthumously in 1631. The *Praxis* (to use the common abbreviation) was a flawed production, imperfectly compiled from a draft in Harriot's hand by his friends Walter Warner and Thomas Aylesbury. It offered what the late Jacqueline Stedall, to whom we dedicate the present volume, described as 'a travesty of his original intentions'.[3] Students of Harriot's involvement with the New World had been hardly better served. His account of the voyage and the 11 months he spent in Virginia, the *Briefe and true report of the new found land of Virginia*, was vivid and informative. But it

---

1 R. Fox (ed.), *Thomas Harriot. An Elizabethan man of science* (Aldershot and Burlington, VT, 2000) and *Thomas Harriot and his world. Mathematics, exploration, and natural philosophy in early modern England* (Farnham and Burlington, VT, 2012).
2 D. B. Quinn, 'Thomas Harriot and the problem of America', in Fox (ed.), *Thomas Harriot. An Elizabethan man of science*, 9–27.
3 J. A. Stedall, 'Reconstructing Thomas Harriot's treatise on equations', in Fox (ed.), *Thomas Harriot and his world*, 53–64 (60).

DOI: 10.4324/9781003096580-1

ran to no more than 48 pages in the original 1588 edition and bore the air of a work written, at least in part, with strategic intent to please the initiator and patron of the expedition, Sir Walter Ralegh.

Today, the sources for the study of Harriot, as well as the secondary literature, are far more plentiful than they were even a quarter of a century ago. But enough was already available in Harriot's and other manuscripts, to allow Stedall to begin her fundamental reappraisal of his contribution to algebra and painstakingly reconstitute the unpublished treatise in which Harriot significantly advanced the understanding of polynomial equations.[4] We can now also draw on a new perspective on the *Praxis*, thanks to a first English translation and scholarly commentary by Muriel Seltman and Robert Goulding.[5] And historians are already benefiting from the long-term international collaboration between scholars at Berlin's Max-Planck-Institut für Wissenschaftsgeschichte and the Universities of Notre Dame and Oxford, which aims to order, digitize, and make publicly available all of Harriot's surviving manuscripts, more than 7,000 pages of them.[6] An early fruit of Matthias Schemmel's involvement in the project has been his reconstruction of Harriot's views on free fall and projectile trajectories, previously known only through the scattered manuscripts but which, through his work, have come to be seen as illuminating the passage between preclassical and classical mechanics.[7]

With regard to America, the *Briefe and true report* remains the essential resource. But we have learned more about the context and publishing history of the various editions of the work, in particular the 1590 edition by the Flemish-born engraver Theodor de Bry. It was in the four versions of this edition, in Latin, German, French, and English, that the fine copperplate engravings, based on the watercolours made on site by John White in 1585–86, were added to the original text, along with detailed commentaries by Harriot. Since then, the illustrations have become inseparable from Harriot's account and done much to perpetuate memory of his role in the early exploration of America. The frequency with which they are reproduced

---

4 J. A. Stedall, *A discourse concerning algebra. English algebra to 1685* (Oxford, 2002) and *The greate invention of algebra. Thomas Harriot's treatise on equations* (Oxford, 2003).

5 *Thomas Harriot's Artis analyticae praxis. An English translation with commentary*, ed. Muriel Seltman and Robert Goulding (New York, 2007).

6 The project, led by Matthias Schemmel in Berlin, Robert Goulding at Notre Dame, and (until her death in 2014) Jacqueline Stedall in Oxford, is described on the ECHO (European Cultural Heritage Online) site at https://echo.mpiwg-berlin.mpg.de/content/scientific_revolution/harriot/harriot_manuscripts. A National Endowment for the Humanities award, announced in January 2022, now supports the continued upgrading of the digitization programmes of the Notre Dame and Oxford teams.

7 M. Schemmel, *The English Galileo. Thomas Harriot's work on motion as an example of preclassical mechanics*, 2 vols. (Dordrecht, 2008). See also Schemmel's 2004 lecture, published as 'Thomas Harriot as an English Galileo: the force of shared knowledge in early modern mechanics', in Fox (ed.), *Thomas Harriot and his world*, 89–111.

and their prominence as the first set of plates in the recent Taschen edition of images from the series of accounts of America that de Bry reissued between 1590 and 1602, bear witness to the undiminished interest in them, not just for their decorative value but more importantly for their historical and ethnographic significance.[8]

The cumulative effect of new resources and the research they have stimulated has been to broaden our perspective on Harriot and his place in the many worlds he touched. Reviewing the first 25 years of the Harriot lectures (in his lecture of 2015, revised for this volume), Stephen Clucas insists on the importance of casting our understanding of Harriot in a broad setting, one that brings out the richness of the diffused common ground – what Schemmel refers to as the 'shared knowledge' – on which he drew and to which he contributed. Ancient texts, the writings of medieval *calculatores*, European innovations with their roots in the work of François Viète, and the experience of contemporary mathematical practitioners were all part of Harriot's rich intellectual armoury. Clucas's point is that, considerable though Harriot's personal legacy was, we can only appreciate its full significance when we set it in the context of the mathematics of his day.

In this volume, Philip Beeley too makes a forceful case for contextualization. Using previously unpublished material, he seeks to broaden our understanding of Harriot's achievements through a study of key contemporaries and successors in the vibrant mathematical community that came to maturity in England during the second half of the seventeenth century. In those years and that community, Harriot's posthumous reputation stood high. The eminent Oxford mathematician John Wallis drew heavily on Harriot's algebra in his own *Treatise of algebra* (1685), and he was not alone in seeing Harriot as an exemplar of the modernity that had come to characterize English mathematics since the early seventeenth century.[9] More than 60 years after Harriot's death (in 1621), though, the story could not be only of him. Setting Harriot in context, Beeley shows how in different ways Nathanial Torporley, Robert Hues, and the two figures responsible for assembling the *Praxis*, Warner and Aylesbury, all had crucial roles in keeping the flame of Harriot's achievements alive.

---

8 De Bry's 1590 edition of the *Briefe and true report* was the first in the America series, which extended by 1602 to fourteen volumes. See M. Van Groesen and L. E. Tise (eds.), *Theodore de Bry. America. The complete plates 1590–1602* (Cologne, 2019) and Larry Tise's contribution to this volume, in particular on the common contemporary practice of colouring the images. On White's watercolours, see Kim Sloan, with contributions by Joyce E. Chaplin, Christian F. Feest, and Ute Kuhlmann, *A new world. England's first view of America* (London, 2007), published on the occasion of an exhibition of the watercolours held at the British Museum in 2007 and subsequently at venues in North America.

9 J. Wallis, *A treatise of algebra, both historical and practical. Shewing, the original, progress, and advancement thereof, from time to time; and by what steps it hath attained to the heighth at which now it is* (London, 1685).

As Beeley argues, the survival of an appreciation of Harriot's work remained precarious even after Wallis's powerful endorsement. Already, following the publication of the *Praxis*, the mathematical papers had risked culling and destruction. In a letter to John Aubrey in 1684 Wallis himself regretted that most of them had been lost and that evidence of the roots of the distinguished mathematical tradition to which he attached such importance hung in the balance. Happily, he was wrong. Passing obscurely and untidily through various hands, the papers – the ones now being ordered and studied in the ongoing international digitization project – had largely survived, as Harriot always hoped they would. Many found their way to Petworth House, the home of George Wyndham, third Earl of Egremont and a direct descendant of Henry Percy, ninth Earl of Northumberland, who had succeeded his friend Ralegh as Harriot's patron in the 1590s. It was at Petworth that the German astronomer Baron Franz Xaver von Zach discovered them in 1784. The discovery resurrected the papers from the status of 'waste' and prepared for the transfer of most of them, by Wyndham, to the British Museum in 1810, leaving the astronomical manuscripts at Petworth.

The role of Wyndham in the preservation of the papers is just one facet of a long and crucial association with the Percy family.[10] The patronage Harriot received, first from Ralegh and then from Percy, exposed him to two different, though overlapping worlds. Ralegh's interest in Harriot was initially of a practical nature; two or three years after taking his BA degree at Oxford, Harriot moved to Durham House, Ralegh's London home on the Strand, to offer instruction in navigational techniques to mariners preparing for voyages to the New World. By contrast, Harriot's contact with Percy gave him the leisure and material security that he used to pursue not only his mathematical interests but also his astronomical observations (with a 'perspective glass' from July 1609, several months before Galileo) and studies of refraction and other optical phenomena that led him to the sine law of refraction 20 years before Willebrord Snell.

Diverse though the priorities of Harriot's patrons were, important continuities between what can too easily be seen as unrelated phases of his intellectual life did exist. Countering the idea of wholly separate realms, David Harris Sacks finds an unbroken thread in Harriot's quest for certainty, whether in mathematics and science or in his exploration of Virginia. He associates the quest with Harriot's religiously inspired belief in a divine *cosmos* whose truths it was his duty to uncover through the active exercise of reason and the senses. Amid the religious and political conflicts of his time, the goal of certain knowledge went hand in hand with an irenicism in the Erasmian humanist tradition that sought peace within the sixteenth-century

---

10 It is an association that continues today in the interest that Max, Lord Egremont has taken for many years in the Thomas Harriot lectures in Oriel and the care of the papers in the Petworth House Archives.

Church, painfully disunited by confessional division and bloody confrontations. Sacks's interpretation offers an alternative to the commonly held view that Harriot kept science and religion apart at a time when suspicions of Ralegh's supposed atheism and Percy's closeness to the circle (including his cousin Thomas Percy) that hatched the Gunpowder Plot of 1605 might have made separation a judicious strategy.

For Sacks, in fact, Harriot's engagement in the discovery of the world and the replacement of doubt and scepticism with the bedrock of certainty, far from being divorced from his religious preoccupations, cannot be understood without them. And he finds the same formative power of religion at work in Harriot's older contemporary, the Church of England clergyman, geographer, and promoter of voyages of discovery Richard Hakluyt. Like Harriot, though some years earlier, Hakluyt owed much to his experience of Oxford, first as an undergraduate at Christ Church and then as a 'student' (fellow) of the college. By the time Harriot graduated, in 1580, Hakluyt had become a prominent figure in the university, mainly through the public lectures he offered on geography. It is likely that Harriot attended these as an undergraduate, and it may even have been Hakluyt who introduced him to Ralegh.

Daniel Carey, too, insists on what serious attention to Hakluyt can add to our understanding of Harriot. In his study of the publishing history of the *Briefe and true report*, he argues for a new focus on the reprinting of the text in the compendium of accounts of English maritime exploration that Hakluyt published as *Principal navigations, voyages, and discoveries of the English nation* in 1589 and again in the enlarged edition of the *Principal navigations* in 1600.[11] In these later printings, the *Briefe and true report* was 'embedded' (Carey's word) alongside other perspectives on the early attempts to establish a colony in Virginia. The result, as Carey shows, is an enriched understanding that serves to bring out the distinctiveness of Harriot's testimony, which is easily lost if the *Briefe and true report* is read independently of other narratives, each of which had its own aims and character. The first account in the *Principal navigations* was that of Arthur Barlowe, a captain on the exploratory voyage to the Carolina coast in 1584; it had an optimistic tone appropriate for what was clearly intended as a promotional document for the eyes of Ralegh, who had sponsored the expedition. The two accounts that followed, by Ralph Lane, a participant in the voyage of 1585–86 and the intended governor of the new colony, were

---

11 Hakluyt, *Principall navigations, voiages, and discoveries of the English nation* (1589). A project, for a 14-volume critical edition of the three massive volumes of the second edition of the work (1598–1600), has been launched under the general editorship of Daniel Carey and Claire Jowitt. Covering English colonial and commercial ventures across the globe, the edition is to be published by Oxford University Press. See the Hakluyt Edition Project site at https://www.hakluyt.org/about/.

distinctly gloomier. They made a nod to the land's commercial potential but also conveyed signs of Lane's anxiety at the instability of relations with the Algonquian people, a condition to which his own aggressive behaviour had undoubtedly contributed. Journals kept by the artist John White, who returned to Virginia in 1587 (to establish the unsuccessful 'lost colony') and 1590, similarly bear witness to the fragility of the colonists' foothold and the poignancy and eventual disillusionment that went with what had come to be regarded, by the time of White's last visit, as a failed enterprise.[12]

How, then, might we reread the *Briefe and true report* in the light of these alternative narratives? Carey's suggestion is that they add significance to the 'slanderous and shameful' comments that clearly troubled Harriot in the wake of the expedition. These seem to have been put about by some who had sailed with Harriot and then, long before the tense winter of 1585–86, gone back to England with the expedition's leader Sir Richard Grenville, evidently disgruntled.[13] Harriot's tone suggests that he took the disparagement personally, as a dismissal of the evidence he had so carefully assembled to promote the prospects for colonization in Virginia. Crucially, as Carey argues, he felt his truthfulness to be in question, and that called for a response. In answering the naysayers, as he did in the prefatory remarks added to the 1590 de Bry edition, he can be seen as affirming the same commitment to a life-long quest for certain knowledge to which Sacks also refers in his chapter.

Despite Harriot's advocacy, the fact remains that the establishment of a viable colony in the area he mapped and explored had to await the movement of settlers from the venture begun at Jamestown to the north in 1607. It was even then a long and tortuous process, only completed when North Carolina was designated a royal colony in 1729. Could it be that Harriot underestimated the difficulty of the venture he was involved in promoting as a 25-year-old in the 1580s? Felipe Fernández-Armesto's core argument is that this was so. Yet how could Harriot's expectations have been so wide of the mark, given the rigours of his own eleven-month stay in the New World? In answer, Fernández-Armesto invites us to look more carefully at Harriot's view of the Spanish experience of empire-building in the Americas.

Inevitably, Harriot gained most of what he knew of the Spain's conquest of the Aztec and Inca empires from the invaders' self-serving tales of the inexorable defeat of indigenous peoples in the face of towering European

---

12 White, appointed governor of the new colony, had accompanied the party of over a hundred settlers in the spring of 1587 but returned to England later in the year to seek supplies. By the time he got back to the site in 1590, the colonists (including his granddaughter Virginia Dare) had disappeared. Since 1937, annual performances of the outdoor drama 'The Lost Colony' have commemorated the events on Roanoke Island.

13 For evidence of Harriot's resentment at the comments, see his pointedly entitled preface, 'To the adventurers, favorers, and wellwillers of the enterprise for the inhabiting and planting in Virginia', in the English 1590 de Bry edition of the *Briefe and true report*, 5–6.

superiority in technology and science, allied to the justifying power of faith and the debilitating effects of native superstition and disease. Such accounts, many of them tinged with the literary tropes of chivalric romance, had little to say about missteps or local resistance. The tone was rather of serene inevitability, and it is an echo of this that Fernández-Armesto sees in the optimism of the *Briefe and true report*. The reality, as he argues, is that the founding and sustaining of an empire in Harriot's day was no easy matter, especially in terrain as unfamiliar and remote from the seat of power as the New World. Failure, in fact, was common, and success less so than we might imagine. Above all, we should be suspicious of the conquistadors' carefully constructed narratives, written with the primary purpose of earning favour with the crown back in Europe. Fernández-Armesto's suggestion is that we temper any assumption of the intruders' superiority and the ease of their victories with a view of the colonizing process through the eyes of the peoples whose lands they colonized.

As Fernández-Armesto observes, many such peoples, especially in the Americas, customarily received strangers with honour; this was what Fernández-Armesto calls the 'stranger-effect'. It was something that Harriot witnessed in his early dealings with the Alongquians of Roanoke and that Spanish colonizers often interpreted as their being received as gods. Crucial in indigenous perceptions of newcomers, however, was also a rational assessment of interests that historians too easily overlook. Powerful invaders could be useful as arbiters in local disputes or as allies in conflicts with neighbours, which meant that deference could make sense. Where the goodwill was maintained on the foundation of a community of interest, an imperial venture could achieve stability. But where colonizers' excessive demands (commonly for food) or harsh responses to petty misdemeanours upset the balance of trust, the upper hand was likely to lie with those who had once welcomed the newcomers. It was a lesson the Algonquians delivered unequivocally as worsening relations, epitomized in the beheading of the local chief Wingina, moved towards the enforced withdrawal of the English settlers in June 1586.

There is every reason to think that Harriot distanced himself from the arrogance and cruelty of the expedition's military leaders responsible for the Wingina affair. What most obviously set him apart was a fascination with Algonquian people and culture. This ran deep, to the point that Mark Horton presents him as an ethnographer long before ethnography is conventionally thought to have emerged, in eighteenth-century Germany. The fact that Harriot went to such pains to learn the Algonquian language, initially from two men from the area who accompanied the exploratory voyage back to England in 1584, is a measure of his determination to come close to a society that so intrigued him. His linguistic competence paid off handsomely in the detailed inside knowledge, extending to the liberal use of local names, that helps to make the *Briefe and true report* such a valuable source for our knowledge of the Indian nations of the Atlantic coast. In this, the modest original edition of 1588, with its six pages on 'the nature and manners of the people', would have been important enough. But, as Horton argues, it was the addition of White's drawings and Harriot's commentaries on them in the de Bry edition of 1590 that gave free rein to Harriot's 'scientific and ethnographic curiosity' and lent force to his description of a people that in certain respects surpassed their English visitors.

Larry Tise's study of the illustrations and commentaries in Harriot's 'magnificent book' similarly insists on the scientific tone of a work that set new standards as an exploration narrative and became one of the most widely distributed books of the age. Like Horton, he notes the careful preparation and the mixture of learning and practical knowhow that lay behind Harriot's systematic account of food and other commodities and the compilation of three remarkable maps of the Carolina coast, two of them surviving as watercolour drawings by White, the other known only as an engraving in the *Briefe and true report*. And he too sees the images and accompanying texts as essential to the lure the book has had for scholars, bibliophiles, and print collectors for over four centuries. Yet, as Tise shows, de Bry's original plan was for an edition in a single language, Latin, to which he intended to add engravings of watercolours done by the artist and cartographer Jacques Le Moyne de Morgues, who had been on an unsuccessful Huguenot-led expedition to Florida in the 1560s. It seems that pressure from Hakluyt may have helped to persuade de Bry to use White's drawings instead. Le Moyne's images were left to appear one year after White's, when de Bry published them with the account of the French colonial venture in Florida by René de Laudonnière.[14]

The passage from drawing or watercolour to copper-engraving and, in many cases, hand colouring involves its own complexities, and the illustrations in de Bry's editions of the *Briefe and true report* were no exception. Tise points to the marked departures from White's original representations. In some cases, engravers introduced embellishments, in the form of backgrounds or decorative elements from White's sketches of animals or plants. More significant was their tendency to transform White's generally expressionless human figures into livelier, more robust beings, and it was the rounded, muscular figures of de Bry's plates that went on to frame perceptions of the indigenous peoples of America for generations to come. These were precisely the kind of people with whom the likes of Ralegh and other European projectors might in due course do business, and it was intended that they should be seen as such.

It is remarkable that a work that began as little more than a brochure for semi-private consumption is still such a rich quarry for research. We can only speculate on what more we might have known if the promised longer

14 *Brevis narratio eorum quae in Florida, Americae provincia, Gallis acciderunt, secunda in illam navigatione duce Renato de Laudonnière* (Frankfurt, 1591).

'Chronicle' on the voyage, to which Harriot refers at the very end of the *Briefe and true report*, had ever been completed and published. But the packed programmes of two international conferences on Harriot to mark the 400th anniversary of his death bear witness to the abundance of materials and problems that still invite attention, and do so on an ever-broadening canvas.[15] Writing as an historical archaeologist, for example, Mark Horton describes the ongoing excavations in the area where Harriot landed. The excavated sites at and close to Fort Raleigh on Roanoke Island have presented particular challenges, not least because of coastal erosion. But work continues on Croatoan Island, now Hatteras Island, to which the 'lost colony' of 1587 may have migrated. So far, traces of any Croatoan settlement have proved inconclusive. They nevertheless enrich our understanding of the material and cultural context of Harriot's ethnographic gaze.

Amid the dazzling multiplicity of Harriot's interests, his mathematical legacy remains, as ever, a major focus of attention. Reflecting on opportunities for future work in his 2015 lecture, Stephen Clucas highlights the optical writings (already being studied by Robert Goulding) and (a continuing interest of his own) Harriot's mechanics. With little in Harriot's papers beyond a short treatise on the collision of round bodies to work from, Clucas has pursued much of his research through the writings of contemporary mathematicians, notably Harriot's friend Walter Warner. The discovery of unknown Warner manuscripts in the Northamptonshire Record Office in the 1990s has given new impetus to this approach. The 39 folios on motion throw light not only on Harriot's work on collisions and the interest in the subject within the Northumberland circle but also on Harriot's possible reading of the work of his older contemporary, the Genevan lawyer Michel Varro. If Harriot had indeed read Varro, this would significantly enrich our understanding of Harriot's place in a mathematical tradition in which, in his day, England lagged behind Continental Europe.

While specialized studies have tended to make the running in recent literature on Harriot, it has been encouraging to see his work coming to the attention of a general readership as well. In this, Robyn Arianrhod's *Thomas Harriot: A life in science* has blazed a welcome trail. Arianrhod casts her biography, the first since John Shirley's almost forty years ago, as a carefully researched but popular account. Writing with a primary focus on Harriot's science and mathematics, she draws heavily on the manuscripts to fashion her view of someone who, in science as in everything he did, was both of his time and an exception. Her journey, beginning with a

---

15 The conferences were 'Thomas Harriot in global and local contexts: a quatercentenary conference', hosted by the Warburg Institute, London (9–10 and 16–17 September 2021) and 'Searchers of new horizons', an OBX History Weekend at the Fort Raleigh National Historic Site, on Roanoke Island, NC, two days of which were devoted to Harriot, native Americans, and early colonial ventures in the area, 31 March - 1 April 2022.

reflexion on D. T. Whiteside's judgement that Harriot may have been Britain's greatest mathematical scientist before Newton and her reading of the *Principia*, was long, and her Harriot evolved along the way.[16] But Arianrhod's guiding thread was a view of him that she characterizes as 'forward-looking'. The implication is not that he was in any sense 'modern' but that his mathematical techniques and experimental observations, important and interesting in themselves, can also and quite properly be seen and evaluated as steps on the road to modern science. Arianrhod is ready to evaluate them accordingly, as she does in her book.

Over the many years of her engagement with Harriot, Arianrhod found herself in the company of a private, enigmatic man who, as she puts it, got under her skin. It is an experience that other biographers will recognize. In the end, Harriot 'came alive' for Arianrhod, both as a man with qualities of genius and as someone whose open-mindedness and sympathy illuminated his account of the people of Virginia. Yet, as she insists, her Harriot retains a mystery that means our 'search' for him is far from ended. In this, Arianrhod's conclusion echoes Stephen Clucas's in his Harriot Lecture of 2015. Despite the research of recent years, we can be sure that new Harriots remain to be discovered. And those other Harriots, as Clucas puts it, can be counted upon to ask as yet unformulated questions of students of early modern science for generations to come.

16 D. T. Whiteside, 'In search of Thomas Harriot', *History of science*, 13 (1975), 61–70.

1 The Certain and Full Discovery of the World: Richard Hakluyt and Thomas Harriot

David Harris Sacks[1]

Voyagers

Richard Hakluyt (1552?–1616) and Thomas Harriot (1560–1621) were explorers and discovers in newly opened territories of study. Both men were consumers of learning and producers of knowledge, and both sought the truth insofar as human reason could grasp it. Both also contributed to the formation of the culture of the modern sciences with their distinctive ethos of debate and proof. Although the lives of the two figures had diverged by the beginning of James I's reign, they were closely connected earlier in their careers, almost certainly starting in Oxford in the later 1570s. In the 1580s both were tied to Walter Ralegh's 'Virginia' project on the Atlantic coast of North America. Nothing better illustrates this linkage than the 1590 illustrated edition of Harriot's *Briefe and true report of the new found land of Virginia*, promoted and translated by Hakluyt and edited and published in English, French, German, and Latin by Theodor de Bry in Frankfurt am Main.[2]

Hakluyt, the older man, promoted ventures of overseas discovery and published the results of those who had made them. His own explorations, as he would emphasize, were largely in journeys to libraries and archives. He

1 This essay is a revised and expanded version of the Thomas Harriot Lecture that I delivered in Oriel College, Oxford on 24 May 2014 and reprised in the Early Modern British and Irish History Seminar, Faculty of History, Cambridge University on 4 February 2015. I am grateful to the Provost and Fellows of Oriel College and the convenors of the seminar in Cambridge for their invitations to speak and their generosity, and to the audiences on both occasions for their questions and comments. I also thank the following for their generous advice on specific points and their friendly encouragement: Anne Oravetz Albert, Robert Anderson, Philip Beeley, David Boruchoff, Louis Caron, Stephen Clucas, Faye Getz Cook, Harold Cook, Nandini Das, Robert Fox, Anthony Grafton, Stephen Johnson, Scott Mandelbrote, Anthony Milton, John Morrill, Anthony Payne, William Poole, Richard Serjeantson, Sophie Smith, Alexandra Walsham, and Andrew Zurcher.
2 Thomas Hariot [Harriot], *A briefe and true report of the new found land of Virginia*, ed. Theodor de Bry, trans. R. Hakluyt (Frankfurt, 1590; STC 12786; USTC 511530); for the other 1590 editions, see USTC 10355 (French); USTC 2212853 (German); USTC 609235 (Latin).

DOI: 10.4324/9781003096580-2

made discoveries derived primarily from the sources he collected, edited, and published in his *Principal navigations of the English nation.*[3] Hakluyt was a conduit for knowledge acquired by others, and a promoter and guide to discovery. Harriot made discoveries at first hand, some in Virginia with which Hakluyt also was connected, but most in the realms of mathematics and what we now know as the natural sciences. However, apart from his account of his travels to the 'new found land of Virginia', which was first published without illustrations in 1588, none of his discoveries reached print during his lifetime.[4] Much of what we know about him is the result of painstaking studies of his surviving manuscripts.

Hakluyt was an ordained Church of England priest and licensed preacher, a minister in several parishes, and from 1602 until his death in 1616 an active Canon of Westminster Abbey.[5] In 1601 he also identified himself as one of Sir Robert Cecil's chaplains.[6] He began giving sermons in 1578, regularly identified himself thereafter as a 'preacher', and is buried in Westminster Abbey.[7] At least twice in his life, he was described in official documents as a 'professor of theology'.[8] Although Harriot was a layman who left behind no systematic statement of his religious beliefs, his friends after his decease in 1621 erected an epitaph to him in London's Church of St Christopher le Stock that identified him as a devoted believer in the Trinity, excelling in theology as well as mathematics and philosophy.[9]

3 Two editions of *Principal navigations* were published in Hakluyt's lifetime: Richard Hakluyt, *The principall nauigations, voiages and discoueries of the English nation made by sea or ouer land* (London 1589; STC 12625), cited hereafter as *PN* (1589), and a second edition, which exists in two versions: Richard Hakluyt, *The principall nauigations, voiages and discoueries of the English nation made by sea or ouer land*, 3 vols. (London, 1598–1600; STC 12626; London, 1599–1600, STC 12626a). Subsequent references will be to this second version, cited as *PN* (1599–1600).

4 Harriot, *A briefe and true report* was also printed by Hakluyt in *PN* (1589), 748–74 and again in *PN* (1599–1600), vol. 3, 266–86. Some mathematical discoveries were published posthumously; see Thomas Harriot, *Artis analyticae praxis*, ed. Walter Warner (London, 1631; STC 12784).

5 For a succinct guide to the trajectory of Hakluyt's life and career, see, D. B. and A. M. Quinn, 'A Hakluyt chronology', in *The Hakluyt handbook*, ed. D. B. Quinn, 2 vols. (Hakluyt Soc., 2nd ser., 144–145, 1974), vol. 1, 263–331. See also George Bruner Parks, *Richard Hakluyt and the English Voyages*, ed. James A. Williamson, American Geographical Society, Special Publication No. 10 (New York, 1928); Peter Mancall, *Hakluyt's promise. An Elizabethan's obsession for an English America* (New Haven, CT, 2007).

6 Richard Hakluyt, 'To the Right Honorable, Sir Robert Cecill Knight, principall Secretarie to her Maiestie', dated 29 October 1601; António Galvão, *The discoueries of the world from their originall vnto the yeere of our Lord 1555*, ed. Richard Hakluyt (London, 1601; STC 11543), sig. A2[c]ᵛ.

7 Christ Church [ChCh] Archives, *Christ Church disbursements, 1578–1579*; Ch.Ch. MS xii.b.21, f. 32r; he identified himself as '*verbi Dei Minister*' to Queen Elizabeth I, Richard Hakluyt, *Analysis seu resolutio perpetua in octo libros Polilitcorum Aristotelis*, BL Royal MS 12 G. XIII, 2; he used 'Preacher' in the dedications to Lord Charles Howard, Lord Admiral, in, *PN* (1599–1600), 1:sig *3 v; and Sir Robert Cecil, PN (1599–1600), 2: sig. *4 v, 3:sig (A3)v; 'Preacher' is also used on the titlepages of all three volumes of *PN* (1599–1600).

8 In 1585 and again in 1612; Quinn and Quinn, 'Hakluyt chronology', vol. 1, 288, 327.

9 John W. Shirley, *Thomas Harriot. A biography* (Oxford, 1983), 417.

However, in both cases a number of modern scholars have come to doubt that their religious beliefs, whatever they might have been, formed the basis for their intellectual projects, or significantly influenced their researches and discoveries. Richard Tuck, for example, has argued that the motivations for Hakluyt's project 'on the whole completely eschewed any religious justification'.[10] David Armitage has argued similarly. 'Religion shaped little, if any, of Hakluyt's corpus', he says, 'either generically or rhetorically'.[11] Andrew Fitzmaurice has largely agreed, as has Peter Mancall.[12] Harriot has received comparable treatment. Modern scholars have taken him not just to have been 'more than normally reticent about discussing [his] private religious beliefs', as Shirley has suggested, but in Arianrhod's view, as consciously keeping 'his religious and astrological beliefs separate from the experimental work'.[13]

In this essay I propose to offer an alternative view which treats Hakluyt's and Harriot's quest for knowledge of the world as itself shaped by religion. It sees their intellectual projects in light of the overlapping epistemological crises of their time, generated on the one side by the explosion of learning associated with Renaissance humanism, overseas discovery, and the so-called new sciences, and on the other by the seemingly irreconcilable conflicts about religious truth and the brutal episodes of confessional strife – wars of religion – between Catholics and Protestants set in motion after 1517 by the Reformation movement.[14] Hakluyt called the worst of these events 'Thyestean tragedies', i.e., acts of cannibalistic revenge.[15] These conditions, in their turn, called

---

10 Richard Tuck, *The rights of war and peace. Political thought and the international order from Grotius to Kant* (Oxford, 1999), 110.
11 David Armitage, *The ideological origins of the British Empire* (Cambridge, 2000), 71; see also 76–77, 81, 85.
12 Andrew Fitzmaurice, *Humanism and America. An intellectual history of English colonisation, 1500–1625* (Cambridge, 2003), 50–55, 138–39; Mancall, *Hakluyt's promise*, 5, 7, 72, 196–97.
13 Shirley, *Thomas Harriot*, 58; Robyn Arianrhod, *Thomas Harriot. A life in science* (Oxford, 2019), 24.
14 For the effects of this confessional strife in Oxford and Cambridge, see Mark H. Curtis, *Oxford and Cambridge in transition, 1558–1642. An essay on changing relations between the English universities and English society* (Oxford, 1959), 165–226; Claire Cross, 'Oxford and the Tudor state'; Jennifer Loach, 'Reformation controversies'; and Penry Williams, 'Elizabethan Oxford: state, church and university', in *The history of the University of Oxford. Volume III: The collegiate university*, ed. James McConica (Oxford, 1986), 117–50, 363–440; H. C. Porter, *Reformation and reaction in Tudor Cambridge* (Cambridge, 1958), esp. 1–242; Ceri Law, *Contested reformations in the University of Cambridge, 1535–1584* (London, 2018). See also Victor Morgan, with Christopher Brooke, *A history of the University of Cambridge. Volume II: 1540–1750* (Cambridge, 2004), 99–146, 437–63.
15 Richard Hakluyt, '*Illustri et Magnanimo Viro, Gualtero Ralegho, Equiti Anglo, Cornubiae & Exoniae stannifodinarium, omniúque Regiae maiestatis castellorum in iisdem prouinciis Praefecto Generaii, S.D.*', dated Paris *octavo Kalendas Martij* [i.e. 22 February], 1587, in Pietro Martire d'Anghiera, *De orbe novo decades octo*, ed. Richard Hakluyt (Paris: apud Guillaume Auvray, 1587; USTC 170862); sig. ã[5]v; cited hereafter as Hakluyt, 'Epistle dedicatory to Sir Walter Ralegh'.

into being intellectual, cultural, and institutional strategies aimed to bring a modicum of peace to Christendom. In what follows, I shall consider the intertwining of two approaches in the work and thought of Hakluyt and Harriot: support for religious toleration and appeals to God's Book of Nature, both based on the belief that God had left certain matters of theology hidden until fully revealed at the Endtime.

This essay, then, is about how Harriot and Hakluyt endeavoured to come to terms with doubt *and* certainty or, to be more precise, about how they envisioned the replacement of doubt with certainty. I shall begin with Hakluyt, from whom I have derived the title of this essay.

Hakluyt: certainty and exchange

In his preface to the 1589 edition of his *Principal navigations* Hakluyt distinguished between those 'wearie volumes bearing the titles of vniversall Cosmographie' from the history of travel, which he said would 'bring vs to the certayne and full discouerie of the world', his goal.[16] This kind of knowledge – knowledge of the world – is empirical knowledge, i.e., dependent on direct observation by the senses. In the modern sciences, as they have developed, it depends not on certainty, but on doubt, or rather on the interplay of the one with the other. Observations of the world and experiments testing them are made to overcome doubt. But for an empirical claim to have truth-value, it must be possible to falsify it with further observations and experiments. Truths about the world, then, necessarily lack certainty; they are contingent and provisional, subject to negation by subsequent discoveries.

However, in Hakluyt's and Harriot's day, and long before, a different conception of certainty, Aristotelian in its foundations, prevailed among the learned. In this earlier paradigm, embodied in Oxford's curriculum during Hakluyt's and Harriot's time there, *scientia* – commonly translated as 'knowledge' – referred generically to any organized form of knowledge, i.e., as Gaukroger has emphasized, to 'a systematic and encyclopedic form of the presentation of knowledge in which known facts were grasped in terms of their underlying principles and causes'.[17] Certainty applies in a strict sense only to these speculative disciplines, pursued for their own sakes, from which alone could be gained knowledge of primary causes or first principles

---

16 'Richard Hakluyt to the fauourable Reader', *PN* (1589), sig. *3r. Among the universal cosmographies Hakluyt had in mind almost certainly was André Thevet, *La cosmographie vniverselle*, 2 vols. (Paris, 1575; USTC 1595. 1597. 62801); Hakluyt became acquainted with Thevet, the French royal cosmographer, while serving between 1583 and 1588 as chaplain to Sir Edward Stafford (1552–1605), Queen Elizabeth I's ambassador in Paris; Quinn and Quinn, 'Hakluyt chronology', vol. 1, 281, 288, 292, 294.

17 Stephen Gaukroger, 'The unity of natural philosophy and the end of *scientia*', in *Scientia in early modern philosophy. Seventeenth-century thinkers on demonstrative knowledge from first principles*, ed. Tom Sorrell, G. A. J. Rogers, and Jill Kraye (Dordrecht, 2010), 20.

that always are so. It does not apply to subjects dependent on observation or to practical subjects, those concerned with actions or *praxis*, which result only in probable knowledge. Astronomy, physics, metaphysics, optics, and arithmetic were included among the speculative disciplines, since, like geometry, also included, they depended on demonstration for proof, either by deduction from universals or by induction to first principles. They produced true and certain knowledge – i.e. knowledge of universals that could not be otherwise. In consequence, they could be taught, that is demonstrated in the strict meaning of the term. They had no dependence on observation, let alone on anything like an experiment. In this setting, a teacher's instruction – *eruditio* in Latin – was regarded as the path to *scientia*, since the teacher imparts organized learning to students with as yet unformed minds. That is, the educations they received fundamentally involved disciplining their speculative capacities. Practical knowledge, in disciplines such as moral philosophy or history, was understood to concern particulars and actions that could be otherwise, to rely on rhetoric rather than logic, and to proceed by probable arguments based on opinion or on probable premises based on observations. It led to understandings that were true only for the most part, but not certain. Hence it had to be acquired by training and habit, rather than demonstration.[18]

According to the Aristotelian paradigm, knowledge must be universal and comprehensive. Theology, regarded as the Queen of the Sciences, was the model of such demonstrable knowledge, i.e., knowledge that is proven syllogistically from self-evident first principles. However, as Aquinas emphasized, all human knowing is imperfect and limited, vexed with doubt and uncertainty.[19] The implication, of course, is that as heirs of Adam and Eve, the capacity of the human intellect to gain certainty of knowledge on their own power is tainted by sin. In consequence, we must rely for the most part on 'opinion' which combines both truth and falsehood and necessarily is corrupted. Equally to the point, religious beliefs – e.g., about the Trinity – depend on revelation, and are beyond demonstration. Strictly speaking, their truths, although absolutely certain, are not knowledge – not *scientia*.[20] Nevertheless, for Aquinas, in Marcia Colish's account, belief in God is dependent on 'faith'

18 *Nic. Eth.*1.2.1094b23–25, 6.3.1139b14–35, in Aristotle, *Nicomachean Ethics*, trans. W. D. Ross, rev. J. O. Urmson, in *The complete works of Aristotle. The revised Oxford translation,* ed. Jonathan Barnes, 2 vols. (Princeton, NJ, 1984), vol. 2, 1730, 1799; *Anal. Post.* 1.1–2.71a1–72b4, in *Posterior Analytics*, trans. Jonathan Barnes, ibid., vol. 1, 114–16; Sophie Smith, 'The language of "political science" in early modern Europe', *Journal of the history of ideas*, 80:2 (April 2019), 203–26, Richard Serjeantson, 'Proof and persuasion', in *The Cambridge history of science. Volume 3: Early modern science*, ed. Katherine Park and Lorraine Daston (Cambridge, 2008), 136–38.

19 Edmund F. Byrne, *Probability and opinion. A study in the medieval presuppositions of the post-medieval theories of probability* (The Hague, 1968), 56.

20 Ethan H. Shagan, *The birth of modern belief. Faith and judgment from the Middle Ages to the Enlightenment* (Princeton, NJ, 2018), 44–45.

which 'transcends science'.[21] '[I]ts knowledge does not attain the perfection of clear sight' through demonstration, Aquinas says, 'wherein it agrees with doubt, suspicion and opinion'. However, it 'deals with the noblest possible object of knowledge' and to those who have it, it is 'absolutely certain'. To everyone else it 'cannot be verified'.[22] In this model, certainty of belief in God's truths can only be achieved through the divinely ordained and sanctioned authority of the Catholic Church, inspired by the Holy Spirit. When there is doubt about what faith required, it is resolved by obedience to the Church.[23] In contrast, the quest for 'the certayne and full discovery of the world', i.e., the quest for knowledge deemed in the Aristotelian tradition to be probable and contingent, represents an intellectual innovation and a radical break with this intellectual paradigm.

As a sometime Oxford divinity student and learned Christian cosmographer, Hakluyt treated the reunification of the dispersed peoples of the world, scattered after Babel, and the recovery of the knowledge lost by Adam and Eve through at their Fall, as mutually reinforcing, leading inexorably to Christ's Second Coming and the Endtime. In this light, he argued that Ralegh had set in motion a civilizing processes in his 'Virginia' colony, first by drawing the region's indigenous population through beneficial exchange into life in civil society and then by bringing them to Christianity and baptizing them. The move from subduing the native peoples to their conversion, he told Ralegh, would counterbalance Europe's descent into the barbarism of what he called the 'Thyestean tragedies' of its wars of religion. The process, as he envisioned it, assumed, as did Spanish missionaries, that the native peoples of the Americas must be removed from their state of near savagery to civil order before their conversion to Christianity can be completed.[24]

The same providential theme appears in the preface to readers with which Hakluyt introduced the second edition of *Principal navigations*. There he compares his 'traueile and cost' in collecting, from libraries and other depositoriesm the documents and accounts of 'Nauigations by Sea ... voyages by land, and traffiques of merchandise by both' to assemble, using Geographie and Chronologie ... the right eye and left eye of all history', 'ech particular

21 Marcia L. Colish, *The mirror of language. A study of the medieval theory of knowledge*, revised edn. (Lincoln, NB, 1983), 128.

22 Aquinas, *Summa theologica*, IIa–IIae, 2. 3. a.1; St Thomas Aquinas, *Summa theologica*: trans. Fathers of the English Dominican Province, 3 vols. (New York, 1947), vol. 2, 1179–80; Shagan, *Birth of modern belief*, 45–46.

23 St Augustine, *Against the Epistle of Manichaeus, called fundamental*, 5.6, trans. Richard Stotherd, in *Nicene and Post-Nicene Fathers, First series*, ed. Philip Schaff (Buffalo, NY, 1887), vol. 4, 213; Shagan, *Birth of modern belief*, 50–55.

24 Hakluyt, 'Epistle dedicatory to Sir Walter Ralegh', d'Anghiera, *De orbe novo*, sig. ã[5]v: see Anthony Pagden, *The fall of natural man. The American Indian and the origins of comparative ethnology* (Cambridge, 1982), 119–97; David Harris Sacks, 'Discourses of Western planting: Richard Hakluyt and the making of the Atlantic world', in *The Atlantic world and Virginia, 1550–1625*, ed. Peter C. Mancall (Chapel Hill, NC, 2007), 436–46.

relation' into a coherent historical narrative to the restoration of a natural body to its original wholeness.[25] The story, as he tells it, represents humanity's journey away from ignorance towards knowledge in a unified and peaceable world governed by civility and reason.[26]

Hakluyt's description of rescuing hitherto lost 'Antiquities smothered and buried in darke silence' from the 'greedy and devouring jaws of oblivion' drew on the myth of Osiris, the smothered, then dismembered, but ultimately resurrected Egyptian Sun god, to whom the ancients gave credit for ending the savage practice of cannibalism among the peoples of the world and bringing them to civilization. As told by Plutarch and Diodorus Siculus, Osiris's body, torn limb from limb, was dispersed into the Nile by his jealous brother, Typhon, god of destruction, from where Isis, Osiris's wife and sister, goddess of wisdom and knowledge, navigating her way through the marshes to search for and recover the dismembered parts, reunited them, and brought him back to life in body as well as spirit.[27] Critical to Plutarch's version is not only the identification of Osiris and Isis with civility and reason, but also his treatment of their brother Typhon as their opposite, a Satan-like figure of 'ignorance and self-deception', who not only dismembers the sacred body of a god, but 'tears to pieces and scatters to the winds ... sacred writings', which Isis then 'collects and puts together and gives to the keeping of those initiated into the holy rites'.[28]

For Christians, the myth of Isis and Osiris, mocked and condemned by St Augustine, could be used to represent the resurrection of Jesus in his body from the clutches of death.[29] It was depicted that way by Bernardino Pinturrichio in the Borgia Apartments in the Vatican, when he provided paintings for it between 1493 and 1495.[30] Hakluyt follows this interpretation.

---

25 Richard Hakluyt, 'A preface to the reader as touching the Principall Voyages and discourses in this first part', *PN* (1599–1600),vol. 1, sig. *3[b]ʳ; see Nandini Das 'Richard Hakluyt's two Indias: textual *sparagmos* and editorial practice', in *Richard Hakluyt and travel writing in early modern Europe*, ed. Daniel Carey and Claire Jowitt (Farnham, 2012), 127.

26 See David Harris Sacks, 'Rebuilding Solomon's Temple: Richard Hakluyt's Great Instauration', in *New worlds reflected. Travel and utopia in the early modern period*, ed. Chloë Houston (Farnham, 2010), 33–36; *idem*, 'To heal the world: commercial exchange as a form of friendship in Renaissance thought', in *Friendship and sociability in premodern Europe. Contexts, concepts, and expressions*, ed. Amyrose McCue Gill and Sarah Rolfe Prodan (Toronto, 2014), 294–97.

27 Plutarch, 'Isis and Osiris', *Moralia* 351C–384C, in Plutarch, *Moralia, with an English translation*, ed. and trans. Frank Cole Babbitt, 15 vols. in 16 (Cambridge, MA, 1927–69), vol. 5, 6–191; Diodorus Siculus, *Diodorus of Sicily, with an English translation*, ed. and trans. C. H. Oldfather, 12 vols. (London and New York, 1933–67), Book I, 11.1–6, and 14.1–27.6, in vol. 1, 36–41, 47–91.

28 Plutarch, 'Isis and Osiris', *Moralia* 352 A, 5:8–9.

29 Augustine, *The City of God against the pagans*, ed. and trans. R. W. Dyson (Cambridge, 1998), book 6, ch. 10, 262; bk. 8, ch. 27, 356-58; bk. 10, ch. 11, 409; see also Juan Luis Vives's commentary on book 2, ch. 14 in St Augustine, *Of the Citie of God. With the learned comments of Io, Lod. Vives*, trans. J[ohn] H[ealy] (London, 1610; STC 916), 74–76.

30 Brian Curran, *The Egyptian renaissance. The afterlife of ancient Egypt in early modern Italy*

Just as Plutarch likens Isis's recovery, reassembly, and revivification of the body of Osiris to the restoration of a sacred text, Hakluyt's appropriation of the figure in effect equates the discoveries of the navigators with the work of Isis in locating Osiris's dismembered limbs, reassembling them, and restoring them to life, with his own recovery of the history of English navigation. That is, in his use of the Osiris story, Hakluyt identifies himself both with the navigators who in making new discoveries were, in effect, reassembling the world as a body, and with Isis, the spirit of reason, who not only reassembled the dismembered Osiris and restored him to life but also collected and put together 'sacred writings' to present to those initiated into 'holy rites'. In thus performing 'stern and rigorous services in shrines', she is said to have sought 'knowledge of Him who is the First, Lord of All, the Ideal One'. For Hakluyt, as a clergyman performing the Church of England 'stern and rigorous services', bringing *Principal Navigations* to print itself represented an act of devotion or worship.[31]

When and where did Hakluyt receive this calling? As he told Sir Francis Walsingham, in dedicating the first edition of *Principal navigations* to him, the moment came while he was still a pupil at Westminster School – in 1568 or 1569 – when he made a visit to his older cousin Richard in his Middle Temple chambers. There, finding 'lying open vpon his boord certeine bookes of Cosmographie with an vniversall Mappe', his cousin,

> seeing me somewhat curious in the view thereof, began to instruct my ignorance, by shewing me the diuision of the earth into three parts after the olde account, and then according to the latter, & better distribution, into more: he pointed with his wand to all the knowen Seas, Gulfs, Bayes, Straights, Capes, Riuers, Empires, Kingdomes, Dukedomes, and Territories of ech part ... .[32]

The elder Hakluyt referred in the first instance to the tripartite worldview as shown, likely, in a Ptolemaic world map printed in one of the cosmographies he owned.[33] The new worldview, represented on a large world map owned by the elder Hakluyt, took into account the discoveries in the

(Chicago, 2007), 107–21; Fritz Saxl, 'The Appartamento Borgia', in *idem, Lectures*, 2 vols. (London, 1957), vol. 1, 174–88; vol. 2, plates, 115–24.

31 Plutarch, 'Isis and Osiris', *Moralia*, 352 A, 5:8–9; for Hakluyt's providentialism, see David A. Boruchoff, 'The politics of providence: history and empire in the writings of Pietro Martire, Richard Eden, and Richard Hakluyt', in *Material and symbolic circulation between Spain and England, 1554–1604*, ed. Anne J. Cruz (Aldershot, 2008), 115–17.

32 Richard Hakluyt, 'To the Right Honorable Sir Francis Walsingham', *PN* (1589), sig. *2r. I have discussed this episode on several earlier occasions, most recently in David Harris Sacks, '"To Winne Them by Fayre Meanes": the ethics of exchange in the making of the early English Atlantic', in *Market ethics and practices, c. 1300–1600*, ed. Simon Middleton and James E. Shaw (London, 2018), 204–6.

33 See Valerie Flint, *The imaginative landscape of Christopher Columbus* (Princeton, NJ, 1992), 3–41, esp. plates 1–6.

Americas and elsewhere.[34] Reviewing this modern map, the elder Hakluyt emphasized the 'speciall commodities, & particular wants' of each territory, 'which by the benefit of traffike, & entercourse of merchants, are plentifully supplied'. The list, linking geographical features associated with enterprise with the forms of political regime within which commercial exchange might occur, demonstrated that God had so disposed the world to balance scarcities in one place with abundance in another to promote exchange between regions.[35] The new map, then, revealed that history was now not just unveiling more and more of the truths of God's creation, but bringing the dispersed peoples of the world into mutually beneficial exchange with one another.

By the time the elder Hakluyt offered his instruction to his young cousin and ward, the view that God had placed the things 'necessary for ye vse of men in diuers lands & sundry countries' so that 'al kindes of men shuld be knit together in vnity & loue' had become a commonplace.[36] It is traceable to the ancient Greeks and early Christian Fathers and repeated in diplomatic correspondence and the writings of a number of early modern commentators including Sir Thomas Smith, Jean Bodin, Giovanni Botero, and Hugo Grotius; it would have an equally impressive history of citation in the seventeenth and eighteenth centuries.[37] However, the most important source conveying this ancient view into the Renaissance and beyond was Desiderius Erasmus's frequently republished *Querela pacis* – 'The Complaint of Peace' – which first appeared in 1517.[38] In this work the humanist, speaking in the voice of 'Peace', offers the 'reasons Nature has provided for concord'. Nature 'was not satisfied', the Goddess says,

34 The map probably was Abraham Ortelius's cordiform projection of 1564: Abraham Ortelius, *Nova totius terrarum orbis iuxta neo tericorum traditiones descriptio* (Antwerp, 1564). See Richard Hakluyt, the elder to Abraham Ortelius, c.1567–68, in *The original writings and correspondence of the two Richard Hakluyts*, ed. E. G. R. Taylor, 2 vols. (Hakluyt Soc. 2nd ser., 76–77, 1935), vol. 1, 77–78n and 76–83 *passim*; R. A. Skelton, 'Hakluyt's maps', in *Hakluyt handbook*, vol. 1, 48–49.

35 Hakluyt, 'To the Right Honorable Sir Francis Walsingham', *PN* (1589), sig. *2r.

36 [Thomas Becon], *The flour of godly praiers* (London, 1550; STC 1719.5), f. 30r-v.

37 See, for example, Jacob Viner, *The role of providence in the social order. An essay in intellectual history* (Princeton, NJ, 1972), 27–54; Douglas A. Irwin, *Against the tide. An intellectual history of free trade* (Princeton, NJ, 1996), 11–25; David Harris Sacks, 'The true temper of empire: dominion, friendship and exchange in the English Atlantic. c. 1575–1625', *Renaissance studies,* 26 (September 2012), 534–40; *idem*, 'The blessings of exchange in the making of the early English Atlantic', in *Religion and trade. Cross-cultural exchanges in world history, 1000–1900*, ed. Francesca Trivellato, Leor Halevi, and Cátia Antunes (Oxford, 2014), 76–82.

38 Desiderius Erasmus, *Querela pacis undique gentium eiectae profligataeque* (Basel, [apud Johann Froben, 1517]; USTC 689404); *idem*, *Querela pacis undique gentium ejectae profligatiaeque* [A Complaint of Peace Spurned and Rejected by the Whole World], trans. Betty Radice, in *The collected works of Erasmus*, vol. 27, ed. A. H. T. Levi (Toronto, Buffalo, London: University of Toronto Press, 1986), 292–322.

simply with the attractions of mutual good will; she wanted friendship to be not only enjoyable but also essential. So she shared out the gifts of mind and body in a way that would ensure that no one should be provided with everything and not need on occasion the assistance of the lowly; she gave men different and unequal capacities, so that their inequality could be evened out by mutual friendships. Different regions provide different products, the very advantage of which taught exchange between them ... Need created cities, need taught the value of alliance between them ...[39]

By the sixteenth century this view was widely accepted as an intrinsic feature of the world as God had created it. It is captured, for example, in the epigram that John Wheeler, sometime Secretary of the Society of Merchant Adventurers of England, used to frame the text of his *Treatise of commerce* in 1601: 'By commerce peoples widely separated by sea and mountains are brought together, so that whatever is produced anywhere is distributed to all'.[40] As Wheeler put it: 'There is nothing in the world so ordinarie and naturall unto men, as to contract, truck, merchandise, and traffique with one another' for their respective benefits.[41]

However, the elder Hakluyt's instruction, providential in its own right, did not stop there, 'From the Mappe he brought me to the Bible', the younger Hakluyt says, 'and turning to the 107 Psalme, directed mee to 23 & 24 verses, where I read that they which go downe to the sea in ships, and occupy by the great waters, they see the works of the Lord, and his woonders in the deepe'. As Hakluyt retold the tale, this move from map to Bible was life-determining: 'The words of the Prophet together with my cousins discourse', he says, 'tooke in me so deepe an impression, that I constantly resolued ... I would by Gods assistance prosecute that knowledge and kinde of literature, the doores whereof (after a sort) were so happily opened to me'.[42] Hakluyt's project became his calling. He represents it as a mission given to him in his cousin's study by God's command. Mancall says that Hakluyt 'made' this pledge as his 'promise to himself'.[43] More importantly, it was also his vow to God.

The verses his cousin chose from Psalms make the point allusively. The text from Psalm 107 as quoted by Hakluyt comes from the Geneva Bible, where – in the accompanying marginal note – safe passage on the high seas is connected with God's love and the redemption from the wages of sin that

39 Erasmus, *Querela pacis*, trans. Radice, 295.
40 '*Commercio Gentes mare, montibusque, discretae miscentur, vt quod vsqum nascitue, apud omnes affluat*'; John Wheeler, *A treatise of commerce* (Middleburgh, 1601; STC 25330), p. 1.
41 Ibid., p. 2.
42 Hakluyt, 'To the Right Honorable Sir Francis Walsingham', *PN* (1589), sig. *2r.
43 Mancall, *Hakluyt's promise*, 24.

comes with faith in Him. The note, itself dependent on commentaries by St Augustine and Cassiodorus, reads: 'He sheweth by ye sea what care God hathe ouer man, for in that y^t he deliuereth the*m* from the great dangers of the sea, he deliuereth them, as it were from a thousand deaths'.[44] In their allegorical interpretations Augustine and Cassiodorus, drawing on their reading of *Revelations*, equate going down to the sea with the descent into hell and the wonders of the deep with baptism and the resurrection.[45]

Hakluyt's passage serves to equate the journey from the Old World to the New not just with the journey from ignorance to knowledge, but from sin to salvation. Taken together, these two movements formed the twin pillars of God's plan for the world. They moved human history itself, as it were, from the map to the Bible, from a focus on the material things necessary for survival in this world to a concentration on divine imperatives for deliverance in the next, with experience of the former leading to, and dependent on, knowledge of the latter. Hakluyt believed that God had revealed the existence of the Americas for His own divine purpose. Seen in this light, therefore, the 'discovery' of the New World, which now was drawing the dispersed nations back into mutually beneficial exchange, represented the inception of a new age. If it inculcated a devoted goal in Hakluyt, it was for a Christianized world ready to welcome Christ's Second Coming.[46]

Hakluyt's judgement derived from his belief that God had created the world for His own providential purpose, and had endowed mankind, made in His own image, with the capacities required to discover it completely and with certainty. He signaled this conviction in his Oxford geography lectures by presenting 'the new lately reformed Mappes, Globes, Spheares, and other instruments of this Arte for demonstration in the common schooles'.[47] In the Aristotelian tradition followed in lectures in these Oxford 'schools', 'demonstration' ordinarily referred to the act of establishing the truth of a proposition in logic or mathematics through deduction or by showing that a conclusion is the necessary consequence of accepted axioms. Here, however, Hakluyt, utilizing the instruments he mentions in a performance, i.e., in a practical exhibition witnessed by observers, means showing how the instruments he presented could be used. Introducing his audiences to the art of navigation and its instruments must then have been one of his purposes. Hakluyt's employing 'demonstration' in this sense is a very early instance of this usage. Although the *Oxford English*

44 Psalms 107:23–24, in the Geneva version, 1560 at f. 258r.

45 St Augustine, *Expositions on the Book of Psalms*, ed. A. Cleveland Coxe, in *A select library of Nicene and Post-Nicene Fathers of the Christian Church*, ed. Philip Schall (Grand Rapids, MI, 1974), vol. 8, 534; Cassiodorus, *Explanation of the Psalms*, trans. and ed. P. G. Walsh, 3 vols. (New York, 1991), vol. 3, 89–90.

46 See David Harris Sacks, 'Cosmography's promise and Richard Hakluyt's world', *Early American literature,* 44:1 (2009), 166, 174, and 161–79 *passim.* Cf. Mancall, *Hakluyt's promise,* 40.

47 Hakluyt, 'To the Right Honorable Sir Francis Walsingham', *PN* (1589), sig. *2r.

dictionary gives the date of the first known occurrence as 1742, where it arises in connection with the public viewing of dissections in an anatomy theater, some practical mathematicians in Hakluyt's era used the term in a similar way. For at least a few of them, a 'geometrical demonstration' showed how something was designed and a 'mechanical demonstration' showed how it worked.[48] As with the usage among surgeons, this stressed the certainty of eye-witnessing, of 'seeing is believing' as it were.[49]

For Hakluyt, therefore, the certain and full discovery of the world involved a double process: direct and accurate eye-witnessing of its diverse parts by explorers and navigators, and a 'demonstration' in an account, equated with a sacred text, using geography and chronology to place each observation or event in its 'due time and place'.[50] Mapmakers and globe-makers took responsibility for precisely and accurately locating each discovery until every particular was situated in its proper place. Cosmographers, like Hakluyt himself, had a corresponding goal. They were charged with setting the stories of discovery into an orderly, time-governed narrative, tracing the sequence of events that would advance in practical terms what Hakluyt understood to be their ordained, providential end.[51]

Harriot: seeing truth

For Hakluyt, the 'world' was the earth as depicted on a terrestrial globe or world map. For Thomas Harriot – mathematician, astronomer, ethnographer, linguist, natural philosopher, and natural historian – the 'world' was the divinely created *cosmos*, 'the heaven and the earth' together in the words of *Genesis*, i.e., the universe as a whole – a new usage in the Renaissance.[52] Like Hakluyt, Harriot's life's project was a quest for certainty of knowledge, albeit in his case, as we shall see, in the face of rationally grounded skeptical doubt. Sometime, probably during the later 1590s when he was studying the rates at which falling bodies hit the ground, he wrote down a

---

48   Mathew Baker (1529/30–1613), master shipwright and mathematical practitioner, employed the terms 'geometrical demonstration' and 'mechanical demonstration' in a similar sense; it meant showing how something worked in practice. See Stephen Johnston, 'Making mathematical practice: gentlemen, practitioners and artisans in Elizabethan England', Ph.D. thesis (University of Cambridge, 1994), 141–43, 153, 157.

49   '*Visus fidelior est auditu*; Seeing is beleeuing'; John Clarke, *Paroemiologia Anglo-Latina in usum schlarum concinnata, or Proverbs English, and Latin* (London, 1639; STC 5360), p. 90.

50   Richard Hakluyt, 'A preface to the Reader as touching the principall Voyages and discourses in this first part', *PN* (1599–1600), 1:sig. *[4]r.

51   See, for example, Gerard Mercator, 'The Booke of Creation and the fabrick of the world', in *idem*, *Atlas, or a geographick description of the regions, countries and kingdomes of the world*, trans. Henry Hexham (London, 1641; Wing STC M1728aA), sigs. A2r-Ir.

52   *OED* '*world, n*'. For an early seventeenth-century instance of this double meaning, see Ralegh, *The history of the world* (London, 1614; STC 20627), I.i.1.1: '*That the inuisible God is seene in his creatures*'.

brief aphorism on the otherwise blank verso side of one of the sheets he had used for this purpose: 'The truth when it is seen is knowne without other euidence'.[53] 'Seen' reports a result, a success. Harriot meant truths that are known, as Richard Hooker would put it, with 'the most infallible certaintie which the nature of things can yield'.[54] At the time, Harriot was one of the ninth Earl of Northumberland's pensioners, having earlier been a servant of Walter Ralegh's.[55]

We know very little about Harriot's early life. His matriculation record at age 17 into St Mary Hall, Oxford on 20 December 1577 tells us he was born in 1560 and was the son of a commoner from Oxford. His father's occupation is not given, nor it is stated whether he was alive or deceased at the time.[56] The entry leaves ambiguous whether his family resided in the city of Oxford or the county; Anthony à Wood says it was the former, probably in St Mary's parish.[57] We also know that Harriot, having studied for four years, was granted a grace for his BA on 29 January 1579/80 without having to participate in that year's Lenten graduation exercises.[58] He would then have begun his Oxford education in 1575 at age 15. Since formal matriculation was only required after a student reached age 16, he may have been connected with St Mary Hall beforehand. It is also possible that he was first taught by a tutor with whom he lodged in the city. Either way, he would have gained a good command of Latin before beginning his Oxford studies. According to Wood, he was 'instructed in grammar learning' within the city of Oxford itself.[59] If so, he may have been a pupil in one of the three formal schools connected with Oxford colleges: Magdalen, New College, or Christ Church, in each of which some of the young students also served as boy choristers.[60]

---

53 BL Add. MS 6788, f. 131 v (undated).

54 Richard Hooker, *Of the lawes of ecclesiastical politie* (London, 1617; STC 13716), 73.

55 On Harriot's biography, see John W. Shirley, *Thomas Harriot. A biography* (Oxford, 1983); Robyn Arianrhod, *Thomas Harriot. A life in science* (Oxford, 2019). For my comments on the latter, see my review in *Intellectual history review*, 31:2 (2021), 369–72 [https://www.tandfonline.com/doi/full/10.1080/17496977.2020.1736444].

56 He is designated a 'plebius filius Oxon'; *Register of the University of Oxford, Volume 2: 1571–1622*, ed. Andrew Clark, Part II: *Matriculations of and Subscriptions (Oxford Hist. Soc*, 11, 1887), 79.

57 Anthony à Wood, *Athenae Oxonieneses ... to which are added the Fasti*, new edn., ed. Philip Bliss, 4 vols. (London. 1813), vol. 1, 299–300.

58 Oxford University Archives (OUA), Register of Congregation and Convocation 1564-82 (NEP Supra Register KK), f. 296r.

59 Wood, *Athenae Oxonienses*, vol. 1, 300.

60 Mary D. Lobel, 'Schools', in *The Victoria history of the County of Oxford*, ed. L. F. Salzman, Volume 1 (Oxford, 1939), 457, 472–74; Nesta Selwyn, 'Education', in *A history of the County of Oxford*, ed. Alan Crossley, *Volume IV: The city of Oxford* (Oxford, 1979), 443–44; see also Mary D. Lobel, 'The grammar schools of the medieval university', in *The Victoria history of the County of Oxford*, ed. H. E. Salter and Mary D. Lobel, *Volume 3: The University of Oxford* (Oxford, 1954), 40–43.

The reason given by Convocation for granting Harriot his grace to take his BA before that year's graduation exercises says simply that 'he cannot wait for another Lent'.[61] Ordinarily an entry of this kind indicates that the petitioner had taken up service as a secretary or tutor in a nobleman's or gentleman's household or as a schoolteacher or clerk, a post that drew on the graduate's linguistic and rhetorical skills. Harriot, however, almost certainly was recruited primarily as a mathematician. Already highly regarded as a practitioner, he had joined the service of Walter Ralegh in Durham House in London by the autumn or winter of 1583, or perhaps early in 1584, soon after Ralegh had been granted the use of the site.[62] One of his known duties was to train Ralegh's mariners in the arts of navigation, a service he might have begun soon after receiving his BA to assist the overseas ventures of Sr Humphrey Gilbert, Ralegh's half-brother, who was preparing what would become his last voyage in 1583.[63] From surviving notes for similar lectures that he later delivered, we can see that, among other things, they demonstrated the uses of navigational instruments. Arguably, he was following Hakluyt's model in his Oxford lectures on mathematical geography, which Harriot as a young mathematician is almost certain to have attended.[64] As we have already seen, he formed a more direct link with Hakluyt in connection with Ralegh's Virginia colony and the publication of his *Briefe and true report*.[65]

However, the parallels between Harriot and Hakluyt are only partial. In seeking knowledge of astronomy as well as natural philosophy, natural history, and geography, the scope of Harriot's interests was wider than Hakluyt's, and in consequence resistant to incorporation into a single framework or overarching structure. Hakluyt's methods were primarily descriptive and emphasized eye-witnessing as the path to certainty of knowledge. Harriot, focusing on the heavens as well as the earth, was more eclectic. However, where he could use mathematics, he did so with the intuition, as we might put it, that mathematics was an ontological and metaphysical leveler. By translating phenomena into the common language of measurement, he could put things and processes on the same plane of being, as it were, and analyse them as abstract structures.

---

61 OUA, Register of Congregation and Convocation 1564–82 (NEP Supra Register KK), f. 296r.

62 Shirley, *Thomas Harriot*, 82.

63 Ibid., 75–80, 86–89; see also Arianrhod, *Thomas Harriot*, 27, 53. Harriot almost certainly was employed by Gilbert or Ralegh on 13 March 1581/2, when he stood surety at the Middle Temple for the 40s admission fine of Philip Amadas; *Minutes of Parliament of the Middle Temple*, ed. and trans. John Hutchinson; *Middle Temple records*, ed. Charles Henry Hopwood, 3 vols. (London, 1904), vol. 1, 249–50; *Register of admissions to the Honourable Society of the Middle Temple, from the fifteenth century to 1944*, ed. H. A. C. Sturgess, 3 vols. (London. 1949), vol. 1, 49.

64 BL Add. MS 6788, ff. 476r–91; Shirley, *Thomas Harriot*, 73, 78–79, 86, 88–89, 95; David B. Quinn, 'Thomas Harriot and the New World', in *Thomas Harriot. Renaissance scientist*, ed. John W. Shirley (Oxford, 1974), 36–37; Jon V. Pepper, 'Harriot's earlier work on mathematical navigation: theory and practice', ibid., 57–75; see also Arianrhod, *Thomas Harriot*, 39–46, 63.

65 *Supra* at n. 2.

Wherever he could, he counted, measured, and timed the phenomena that he observed, and where possible, for example in observing the refraction of light, sought to reduce his results to a mathematical equation.[66] His desire to quantify and calculate appears ingrained in his approach to the world, for example in estimating on a very rainy day how much of it would fall if it continued at the same rate for a full 24 hours.[67] It is evident more profoundly in the thought experiment that Harriot conducted to judge the natural limits of population increase resulting from the union of 'one man and one woman' in the 6,000 years since Creation.[68]

As is well known, and as Harriot himself knew, in the 1590s Ralegh, and Harriot with him, gained a reputation for unbelief or worse.[69] Ralegh was accused of maintaining a 'schoole of Atheisme' in which those involved, including Harriot, not only rejected all divinity but, in keeping with charges of Epicureanism, denied the immortality of the soul.[70] Harriot himself was said to have denied 'the resurrection of the bodye' and to have believed human beings existed before Adam and Eve. He was accused of being a 'Jugler', conjurer, necromancer, and Magus, as well as an atheist.[71] Nathaniel Torporley, the clergyman and mathematician who Harriot made his literary executor, found him in religion 'a frail Man ... deserving of reprehension and anathema' for his errors.[72] A few years later, a close reader of John Selden's *Mare Clausum* (1635), noted in the margin of his

66 J. A. Lohne, "Thomas Harriot (1560–1621): the Tycho Brahe of optics', *Centaurus*, 6:2 (June 1959), 112–21; Shirley, *Thomas Harriot*, 383–85; Arianrhod, *Thomas Harriot*, 146–51.

67 BL Add. MS 6788, f. 411r-v; Shirley, *Thomas Harriot*, 82–83.

68 In these calculations, Harriot almost certainly was acting to aid Ralegh in his work on *The history of the world*. The results are found at BL Add. MS 6782, f. 31r-v; B. J. Sokol, 'Thomas Harriot—Sir Walter Ralegh's tutor—on population', *Annals of science*, 31:3 (1974), 205–12; Arianrhod, *Thomas Harriot*, 235–37.

69 Petworth House Archives (PHA), HMC 241 VIb, item 47; David B. Quinn and John W. Shirley, 'A contemporary list of Hariot references', *Renaissance quarterly*, 22:1 (Spring 1969), 19–20, 26; Nicholas Popper, *Walter Ralegh's History of the world and the historical culture of the late Renaissance* (Chicago, 2012), 82–87; see also Shirley, *Thomas Harriot*, 186–87; Arianrhod, *Thomas Harriot*, 129.

70 [Andreas Philopater (i.e. Robert Persons)], *An aduertisement written to a secretary of my L. Treasurers of Ingland, by an Inglishe intelligencer* ([Antwerp, 1592; STC 19885), 18; this is an abbreviated version of Andreas Philopater [Robert Persons], *Elizabethae, Angliae reginae haeresim Calvinianam propugnantis, saevissimum in catholicos sui regni edictum*, (Lyons: apud Jean Didier, 1592; USTC 156831), 28–29; see Arianrhod, *Thomas Harriot*, 128–29.

71 Thomas Nashe, *Pierce Penilese his supplication to the diuill* (London, 1592; STC 19371), f. 8r-v; Thomas Nashe, *Christ's Teares ouer Ierusalem* (London 1593; STC 18366), f. 58r-v; BL Harl. MS 6848, ff. 185r, 191 v; BL Harl. MS 6849, ff. 184r, 185r; John Aubrey, *Brief lives, with an apparatus for the lives of English mathematical writers*, ed. Kate Bennett, 2 vols. (Oxford, 2016), vol. 1, 109.

72 Nathaniel Torporley, *Corrector analyticus*, Lambeth Palace Library, Sion College MS L40.2/E10; in Henry Stevens, *Thomas Hariot, the mathematician, the philosopher, and the scholar* (London 1900), 164, 172; Shirley, *Thomas Harriot*, 5. A transcription of Torporley's full Latin text (with some errors) can be found in James Orchard Halliwell, *A collection of*

copy that Harriot was 'great philosopher and Mathematician, but a bad Christian and Profane Stoick Philosopher'.[73]

It is sometimes asked why Harriot refrained from publication even though he made a number of original discoveries in mathematics, physics, astronomy, and other subjects.[74] Sir William Lower, his good friend, lamented this failure, which he said 'hath robd you' of the glory you are due. 'Doe you not ... starthe', Lower asked, 'to see every day some of your inventions taken from you ... Onlie let this remember you, that it is possible by to[o] much procrastination to be prevented in the honor of some of your rarest inventions and speculation. Let your countrie & frinds injoye the comforts they would have in the true and great honor you would purchase your selfe by publishing some of your choise works'.[75] More recently, it has been suggested that Harriot 'never achieved the fame he deserved for his discoveries', despite being prodded as Lower did, largely because 'he was a perfectionist'.[76] No doubt there is some truth in this claim. However, other reasons for his failing to publish arguably were more important, as Lower himself hinted in saying to Harriot: 'you know best what you have to do'.[77]

Without a patron at Court after 1605, Harriot had no one to protect him should he be summoned before the authorities again as he had been in connection with the Gunpowder Plot, in which royal officials thought the ninth Earl of Northumberland, his patron, was implicated. Starting in 1605, Northumberland was confined in the Tower at the pleasure of the King, where he joined Ralegh, who in 1603 had been convicted of treason in connection with the so-called Main Plot to replace King James I with Lady Arabella Stuart. Harriot, imprisoned for weeks in the Gatehouse of Westminster Abbey where he was interrogated for any possible implication in the Gunpowder treason, had gained his release not only by professing himself 'truly innocent in hart and thought' and 'always of honest conversation and life', but 'neuer any busy medler in matters of state' or 'ambitious for preferments, [b]ut contented with a priuate life for the loue of learning that I might study freely'. In pleading for his freedom from

the letters illustrative of the progress of science in England from the reign of Queen Elizabeth to that of Charles the Second (London, 1841), 109–16.

73 Aubrey, *Brief lives*, vol. 2, 890–91, citing Oxford, Bodleian Library, MS C. 262, 166–67; the reader was Charles Stanhope, second son of Philip, Earl of Chesterfield, who matriculated at Oriel College in 1632, but did not take his BA.

74 See, for example, Arianrhod, *Thomas Harriot*, 4, 100, 161, 173, 179 182, 222, 239.

75 William Lower to Thomas Harriot, Tra'venti [Trefenti, Wales], 6 February 1610/11. The first half of this letter is known from Franz Xavier von Zach, *Monatliche Correspondenz zur Beförderung der Erd-und-Himmels Kunde*, 28 vols. (Gotha, 1800–13), vol. 8, 47; printed in Stephen P. Rigaud, *Supplement to Dr. Bradley's Miscellaneous works with an account of Harriot's astronomical papers* (Oxford, 1833), 43, from which I am quoting. The second half of this letter survives among Harriot's MSS, BL Add. MS 6789, ff. 427r–428r

76 Arianrhod, *Thomas Harriot*, 4, 100.

77 Lower to Harriot, 6 February 1610/11, in Rigaud, *Supplement*, 43.

confinement to continue his 'labour & endeavours', he promised, in effect, to maintain a quiet private life clear of controversies.[78] As he said later in a letter to Johannes Kepler, written in 1608, it was not possible for him in the present to 'philosophize freely', although he hoped God would soon put an end to the hindrances.[79] That moment never came.

Harriot was not an 'atheist' either in its modern meaning, i.e., an unbeliever in the existence of God, or in its early modern one as 'godless' or 'impious', i.e., an unbeliever in Christian doctrine who acted without regard to its precepts.[80] Neither was he a 'Deist', as John Aubrey and Anthony à Wood called him.[81] As was recorded on his epitaph in the church of St Christopher-le-Stocks in Threadneedle Street, London he not only 'cultivated all the sciences' – theology as well as mathematics and philosophy – but was 'a most studious searcher after truth' and 'a most devout worshipper of the Triune God'.[82] Like Montaigne, his philosophical skepticism was tempered by a form of fideism, according to which *religious* certainty could come only from revelation.[83]

As we earlier suggested, Harriot himself understood certainty of knowledge to come from 'seeing'. On the verso of a sheet he had used for a mathematical exercise concerning free-falling bodies he wrote this previously quoted single sentence: 'The truth when it is seen is knowne without other euidence'.[84]

---

78 Thomas Harriot to 'The Lordes of his Majestys most honorable privy councell', Gatehouse, 16 December 1605; Hatfield MS 114/41, HMC, *Salisbury*, 17:554; Shirley, *Thomas Harriot*, 348–49. Harriot's petition asked for release to give him the freedom to produce useful results 'to the good likening & allowance of the state & common weale'; however, he published none of his important discoveries before his death in 1621.

79 Thomas Harriot to Johannes Kepler, Sion near London, 13 July 1608; in Johannes Kepler, *Gesammelte Werke*, Band XVI, ed. Max Caspar (Munich, 1954), 172.

80 On 'unbelief' in the early modern era and its connection with the concept of 'atheism', the starting place remains Lucien Febvre, *The problem of unbelief in the sixteenth century. The religion of Rabelais*, trans. Beatrice Gottlieb (Cambridge, MA, 1982), esp. 131–51; Earnest A. Strathmann, *Sir Walter Ralegh. A study in Elizabethan skepticism* (New York, 1951), 61–97. See also Shagan, *Birth of modern belief*, 95–96, 100–25; Alec Ryrie, *Unbelievers. An emotional history of doubt* (Cambridge, MA, 2019), 7–9, 14–15, 214.

81 Aubrey, *Brief lives*, vol. 1, 678; Wood, *Athenae Oxonienses*, vol. 1, 301.

82 The Church of St Christopher-le-Stocks was destroyed in the Great Fire of London in 1666, but Harriot's Latin epitaph is recorded in John Stow, *The survey of London* (London, 1633; STC 23345), pp. [831]–832, transcribed and translated in Shirley, *Thomas Harriot*, 474.

83 Montaigne, 'An apology for Raymond Sebond', in Michel de Montaigne, *The complete essays*, ed. and trans. M. A. Screech (London, 1991), 489–683; Richard Popkin, *The history of skepticism. From Savonarola to Bayle* (Oxford, 2003), 29, 31–32, 35, 37. 44–63; Terence Penelhum, *God and skepticism. A study in skepticism and fideism* (Dordrecht, 1983), 18–61; Ann Hartle, 'Montaigne and skepticism', in *The Cambridge companion to Montaigne*, ed. Ullrich Langer (Cambridge, 2005), 186, 189, 191; Thomas D. Carroll, 'The traditions of fideism', *Religious studies*, 44:1 (March 2008), 1–22.

84 BL Add. MS 6788, f. 131r-v; Amir R. Alexander, *Geometrical landscapes. The voyages of discovery and the transformation of mathematical practice* (Stanford, CA, 2002), 98. Since neither the obverse nor the reverse of this folio bears any date, it is uncertain whether

The import of this utterance, which derives from use of the word 'seen' in relation to 'knowne', 'euidence' and 'truth', is ambiguous. In Harriot's period, as today, the verb 'to see' refers most commonly to the perceiving of things with the eyes, but it had long also meant 'to apprehend or perceive with the mind', or 'to understand' or 'to recognize'. As St Augustine made clear in his *Confessions*, there is an element of truth in both possibilities. 'Seeing', Augustine says, 'is the property of our eyes. But we also use the word in other senses, when we apply the power of vision to knowledge generally'.[85]

The central issue is certainty. Possible meanings are revealed in a passage in Richard Hooker's *Laws of ecclesiastical polity*, a work that Harriot may have encountered early in his life, but that he acquired for his own library in 1617.[86] '[M]an', Hooker says there,

> desireth euermore to know the Truth according to the most infallible certaintie which the nature of things can yeeld. The greatest assurance generally with all men is, that which we haue by plaine aspect and intuitiue beholding. Where wee cannot attaine vnto this; there what appeareth to bee true by strong and inuincible demonstration, such as it is not by any way possible to be deceiued, thereunto the mind doth necessarily assent, neyther is it in the choice thereof to doe otherwise.[87]

Hooker follows the Aristotelian distinction between empirical and speculative knowledge, although in contrast to Aquinas and his followers, he places truths known by their 'plaine aspect' to the senses ahead of truths established by logical 'demonstration'. Harriot's comment implies the certainty of logical proofs and solutions to mathematical equations as well as of first-hand reports of visual perceptions.

Harriot's views on skeptical doubt provide a context for his aphorism. Among his manuscripts we find the following comment headed '... how

Harriot was simply looking for a blank sheet of paper or seeking to comment on the obverse of the document.

85 *Confessions*, X.xxxv.54; Saint Augustine, *Confessions*, ed. and trans. Henry Chadwick (Oxford, 1992), 211. Hilary Gatti interprets Harriot's aphorism not as referring 'to sight as a sense impression ... but rather to intuition in the Cartesian sense'; see Gatti, 'The natural philosophy of Thomas Harriot', in *Thomas Harriot. An Elizabethan man of science*, ed. Robert Fox (Aldershot and Burlington, VT, 2000), 82. Alexander treats the passage as granting authority to 'personal observation' as opposed to Scholastic methods; see Alexander, *Geometrical landscapes*, 98–104.

86 John Bill to Thomas Harriot, 31 October 1617; PHA, HMC 241/4 f. 9r-v. However, Harriot might already have read it in the ninth Earl of Northumberland's library in Syon House, which held a copy; PHA 5377, *Catalogus librorum bibliothecae Petworthianae, c. 1690*, f. 26 v; see Scott Mandelbrote, 'The religion of Thomas Harriot', in *Thomas Harriot. An Elizabethan man of science*, ed. Fox, 253, 275; see also Hilary Gatti, 'Natural philosophy of Thomas Harriot', ibid., 83.

87 Hooker, *Lawes* (STC 13716), 73.

happy mayst thou be and yet what thing now more impossible' written for an unknown purpose on a torn sheet of paper, probably in the later part of Harriot's life.[88] The passage then goes on:

> Whether there be any true knowledge at all of things or not, yet euery man is opinioned enough to know something and that truly; even those which haue denied all knowledge. Socrates which by oracle was judged the wisest man in his time said he knew nothing. And Aristarchus which said he knew not as much as that, and Pyrrho with his followers called Skeptics which will not affirm or deny any knowledge to be true or false but do still doubt, yet they make their knowledge certain to doubt assuredly. And there was something certain that made Aristarchus to seem so certain in his opinion.[89]

The body of the passage finds a common thread among three forms of philosophical skepticism – Socratic, Academic, and Pyrrhonian – with Aristarchus mistaken for Arcesilaus, founder of the Second or Middle Academy and its form of Academic skepticism. Each professes to know nothing, or nothing with certainty, yet the members of each school know for certain at least one thing –namely their own doubt.[90] In light of the passage's heading, Harriot plausibly can be read to say that while one can achieve a modicum of certainty about the world, assurance in this life about one's ultimate happiness is impossible, since final happiness comes only with one's ultimate fate. As Herodotus reported, Solon said something similar to Croesus.[91] Translated into Christian terms, Harriot's remarks are about the uncertainty that human beings necessarily must have in their present lives in knowing beyond doubt whether their days will end happily with salvation.

---

88 BL Add. MS 6789, f. 460r; the first word in this heading may have been bound into the volume's binding; Gatti plausibly suggests 'Man' as the first word; see Gatti, 'Natural philosophy of Thomas Harriot', 82. Harriot appears to have been reflecting St Augustine's discussion of the 'happy life' in Book X of *Confessions*, X.xx.29–Xxxiii.34, Chadwick ed., 196–200.

89 BL Add. MS 6789, f. 460r.

90 As Gatti has suggested, Harriot used the name 'Aristarchus' to refer to the founder of Academic skepticism, namely Arcesilaus; see Gatti, 'Natural philosophy of Thomas Harriot', 82, also Stephen Clucas, 'Thomas Harriot and the field of knowledge in the English Renaissance', in *Thomas Harriot. An Elizabethan man of science*, ed. Fox, 135. Harriot possibly collapsed Aristarchus of Samos, the ancient Greek astronomer and mathematician, together with Aristarchus of Samothrace, the grammarian and director of the Library of Alexandria who established the texts of Homer's *Iliad* and *Odyssey*. The latter figure had a reputation among the ancients and with Renaissance humanists as a severe critic whose skepticism made him unable to treat verbal or editorial slips with charity; see, for example, Horace, *Ars poetica*, line 450.

91 Herodotus, 1.29–33; Herodotus, *The history*, trans. David Grene (Chicago, 1987), 44–48.

As Gatti has pointed out, Harriot's view about 'doubt' has an obvious similarity to the *cogito* argument made by Descartes in his *Discours de la méthode* of 1637.[92] However, Descartes was mainly concerned to show, as he would put it, that the 'I which is thinking is an immaterial substance with no bodily element' and therefore was ontologically distinct from matter.[93] Harriot had no such philosophical purpose. Instead, he was drawing for a different purpose on a view offered much earlier by St Augustine in several places, including *The City of God*.[94] However, although we know that Harriot had an interest in Augustine's works, he need not have ever read Augustine's comment to come to his view.[95] As Descartes said when challenged on the similarity of his *cogito* argument to Augustine's view, 'it is such a simple and natural thing to infer ... that it could have occurred to any writer'.[96] Whether Descartes was struggling with hyperbolic doubt, Augustine, at the time he wrote his critique of skepticism, was not. Neither was Harriot. Instead, in affirming the truth of at least one certainty, he was establishing a foundation to search for more. His aim was to show the limits of doubt, by distinguishing the kinds of knowledge the acquisition of which is up to us – mathematical and empirical knowledge – from the certainties that depend on divine revelation and God's grace and will.[97] As Hooker said in addressing the question of 'infallible certaintie', when truths from 'plaine aspect' and 'inuincible demonstration' fail, Christians have 'Scripture ... receiued as the Word of God' as 'the strongest proofe of all, for 'which we haue necessary reason', greater than what 'we see with our eyes'.[98]

92 Gatti, 'The natural philosophy of Thomas Harriot', 82; René Descartes, *Discours de la méthode: pour bien conduire sa raison, & chercher la vérité dans les sciences* (Leiden, Joannes Maire, 1637; USTC 1011874).

93 Descartes to Andreas Colvius, 14 November 1640, in René Descartes, *The philosophical writings of Descartes*, trans. John Cottingham, Robert Stoothoff, Dugald Murdoch, and Anthony Kenny, 3 vols. (Cambridge, 1991), vol. 3, 159–60.

94 See St Augustine, *City of God*, XI:26, ed. Dyson, pp. 483–4; St Augustine, *On the Trinity*, XV:12, in Augustine, *On the Trinity. Books 8–15*, ed. Gareth B. Matthews, trans. Stephen McKenna (Cambridge, 2002), 190–93; Christopher Kirwin, 'Augustine against the skeptics', in *The skeptical tradition*, ed. Myles Burnyeat (Berkeley and Los Angeles, CA, 1983), 205–23; Gareth Matthews, *Thought's ego in Augustine and Descartes* (Ithaca, NY, 1992), 11–38.

95 In 1617, Harriot acquired a copy of St Augustine, *Contra secundum Juliani responsionem*, ed. C. Menard (Paris, apud Sébastien Chappelet, 1617; USTC 6016026), PHA, HMC 241/4 f. 9 v; Mandelbrote, 'Religion of Thomas Harriot', 253, 255. In Northumberland's Syon House library, Harriot also had access to the 1586 Lyon edition of Augustine's *Opera* (USTC 142371) PHA 5377, 19 v, and probably to copies of Augustine's *Confessions* and *City of God* in Ralegh's library in the Tower; see Walter Oakeshott, 'Sir Walter Ralegh's library', *The library*, 5th ser., 23 (1968), 297, 311.

96 Descartes to Andreas Colvius, 14 November 1640, in Descartes, *Philosophical writings*, vol. 3, 159.

97 Cf. Gatti, 'Natural philosophy of Thomas Harriot', 82.

98 Hooker, *Lawes* (STC 13716), 73.

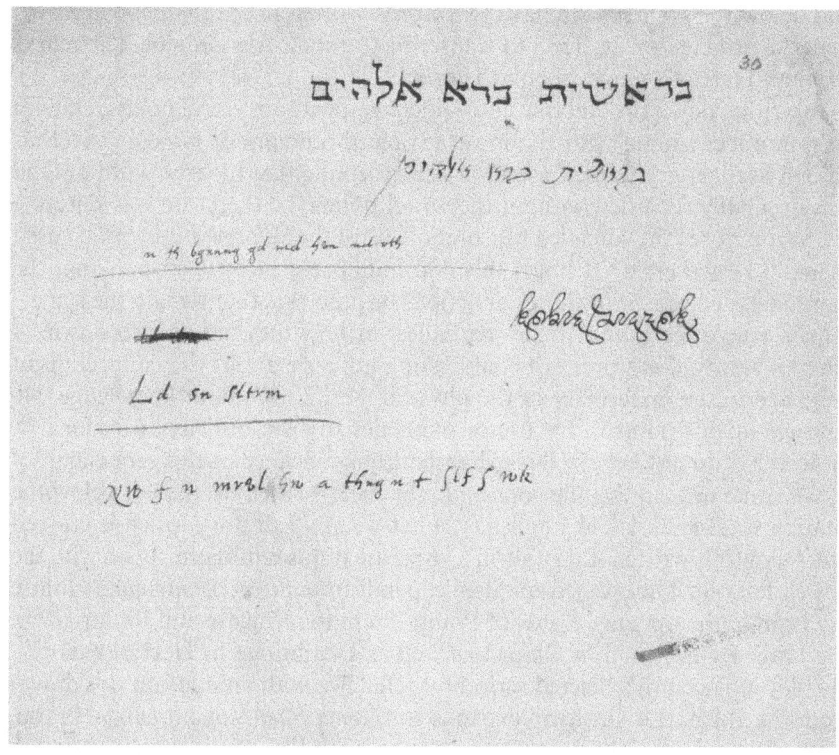

Figure 1.1 Harriot, undated. British Library, Add. MS 6785, f. 30r. By permission of the British Library.

The philosophical and theological balance Harriot sought to strike emerges from a strange document surviving among his manuscripts, almost certainly dating from the last years of Harriot's life as he contemplated his own impending death. It is a meditation on beginnings and endings in which each element must have been written down in sequence over a relatively short period – a few hours at most – as Harriot reflected on what he had just put down on the sheet (see Figure 1.1). Writing in Hebrew, he would have begun on the upper right, where he twice wrote the same three words, once in block Hebrew letters, imitating print, and then in discursive script, each without the 'vowel points', the diacritical marks used to signify vowel sounds: בראשית ברא אלהים (*Bereshit bara Elohim*).[99] These are the first three words of *Genesis* in Hebrew. Next, on the left, are the first words of *Genesis* in English, leaving out the vowels: 'nth bgnnng gd md

99 I am grateful to Anne Oravetz Albert for her assistance with the Hebrew and to her and to Anthony Grafton for their comments and advice.

hvn nd rth', i.e., 'In the beginning God made heaven and earth'. To the right, and below the Hebrew, we have two words written in the phonetic alphabet that Harriot created ca. 1585 to transcribe Carolina Algonquian. The marks spell out Harriot's name: 'Tomas Haryot'.[100]

In writing down the Hebrew in the document without vowel points, Harriot was, in effect, siding with the most advanced scholars of his era – such as Joseph Scaliger and Isaac Casaubon – in accepting that the true word of God was originally recorded without the vowel points; he therefore was rejecting the historicity of the so-called Masoretic Text of the Hebrew Bible, which used them.[101] However, what holds this odd initial sequence together is that the items included represent things or actions or processes that remain the same – remain true or certain or real – regardless of how they are written down.[102] Harriot seems, therefore, to be addressing a puzzle about the connection of language to the materiality of the physical world, not about knowledge. He gestures at this point in his use of phonetics to spell out his own name.[103] However, I do not believe this exhausts the significance of this sequence.

We come next to the allusions and references written on the left below the citation to *Genesis* 1:1 in English.[104] First we see 'Ldt dm', which is crossed out; below it is written 'Ld sn sltrm'. These are phrases in Latin, again with the vowels left out. The crossed-out item expands to '*Laudate Dominum*' – 'Praise the Lord' – the opening words of Psalm 116 in the Vulgate and Psalm 117 in the Latin Psalter used in Oxford as well as Cambridge in Harriot's day.[105] Harriot replaced this deleted scriptural reference with an allusion not drawn from the Bible. 'Ld sn sltrm' expands to '*Lauda Sion Saluatorem*' – 'Praise, Sion, thy Saviour'. These words are the opening line of the sequence dating

100 Vivian Salmon, 'Thomas Harriot (1560–1621) and the English origins of Algonkian lin-guistics', in Salmon, *Language and society in early modern England. Selected essays, 1981–1994*, ed. Konrad Koerner (Amsterdam and Philadelphia, PA, 1996), 143–72. Some of the symbols Harriot used derive from cossic numbers; see Robert Recorde, *The whet-stone of witte* (London, 1557; STC 20820), sig. S1r–Ll2v; Shirley, *Thomas Harriot*, 110–12; Salmon, 'Thomas Harriot', 153.

101 Anthony Grafton, *Joseph Scaliger. A study in the history of classical scholarship*. II: *Historical chronology* (Oxford, 1993), 732–37; Anthony Grafton and Joanna Weinberg, '*I have always loved the Holy Tongue': Isaac Casaubon, the Jews, and a forgotten chapter in Renaissance scholarship* (Cambridge, MA, 2011), 312–13. On the debates about Hebrew vowel points, the writings of Elia Levita are the starting place; see Levita, *The Massoreth ha-massoreth of Elias Levita*, ed. Christian D. Ginsburg (London, 1867) and *Introduction to the Massoretico-Critical edition of the Hebrew Bible* (London, 1897).

102 St Augustine made a similar point in *Confessions*. XI.iii.5, in Chadwick ed., 223–24. on which it seems likely Harriot was reflecting as he composed his meditation.

103 For discussion, see *infra* at n. 143.

104 Here I must acknowledge the invaluable assistance of Andrew Zurcher.

105 Psalm 116:1 in the Vulgate: '*Alleluja. Laudate Dominum, omnes gentes; laudate eum, omnes populi*'. William Byrd provided a setting for it, probably composed in the 1590s: William Byrd, *Gradualia: seu Cantionum sacrarum quarum aliae ad quatuor verò ad quinque et sex voces editae sunt. Liber secundum* (London, 1607; STC 4244.5), sig. H2v–H3r

from 1264 written by St Thomas Aquinas for the Mass for Corpus Christi. Since from the mid-1590s, Harriot lived in a dwelling in Syon Park, just north of the ninth Earl of Northumberland's Syon House, *Lauda Sion* can be read as Harriot's invocation of his own praise to Christ as his saviour. While in Oxford, Harriot might have encountered Psalm 116/117 in the Latin edition of the Psalter used in the University, since the *Book of Common Prayer*, also used in Latin in the University, required it to be read or sung in Morning Prayers on the 24th of each month.[106] When and where Harriot became familiar with *Lauda Sion Salvatorem* is unknown.[107]

We come now to the last line in the sequence, immediately below 'ld sn sltrm'. Again we have a phrase with the vowels, or most of them, left out; the article 'a' has been retained: 'Nw f w mrvl hw a thng n t slf s wk'. Expanded it reads: 'Now if any marvel how a thing in it self so weak'. This fragment comes 600 or so words into the dedicatory letter to Archbishop Whitgift with which Richard Hooker prefaced Book 5 of his *Laws of ecclesiastical polity*. The passage reads as follows:

> Now, if any maruail, how a thing in it selfe so weak, could import any great danger, they must consider not so much how small the spark is that flieth vp, as how apt things about it are to take fire. Bodies politique, being subiect as much as naturall, to dissolution, by diuers meanes, there are vndoubtedly mo[r]e estates ouerthrowne through diseases, bred within themselues then through violence from abroad, because our manner is alwayes to cast a doubtfull and a more suspicious eye towards that, ouer which we know we have least power ... .[108]

---

106 *Psalterium Davidis carmine redditum per Eobanum Hessum* (London, 1575: STC 2356), 268–9; *Liber precvm pvblicavum* (London, 1560; STC 16424), sig. b1r. It is not certain that Harriot regularly attended Morning Prayers while an undergraduate in St Mary Hall, since it lacked its own chapel. It is possible, however, that the students in the Hall participated in Morning Prayers in the chapel of Oriel College next door, or in St Mary's Church across the High Street.

107 The most likely source is the Post-Tridentine *Roman Missal* authorized by Pope Pius V in 1570. *Missale Romanum ex decreto sacrosancti Concilii Tridentini restitutum* (Rome: Bartolomeo Faletti & Giovanni Varisco, 1570; USTC 820648); Pope Clement VIII re-authorized the *Missal* in 1604: *Missale Romanum ex decreto sacrosancti Concilii Tridentini restitutum* (Rome: ex Tipografia Vaticana, 1604, USTC 4030498). For a copy, see *In Sollemnitate Sacratissimi Corporis Christi, Sequentia: 'Lauda Sion saluatorem'*, *Missale Romanum ex decreto Sacrosancti Concilii Tridentini restitutum* (Antwerp: ex officina Christophe Plantin, 1577: USTC 412667), lii. William Byrd provided a setting for *Lauda Sion* in his *Mass Prospers for the Feast of Corpus Christi: Sequentia*, to be sung in connection with Byrd's Mass for four voices for Corpus Christi, first published in 1605; William Byrd, [G]*radualia: ac cantiones sacrae. quinis, quaternis, trinisque vocibus concinnatae* (London, 1605; STC 4243.5).

108 Hooker, *Lawes* (STC 13716), Book 5, sig V3r. The fifth book of Hooker's *Lawes* was first separately published in 1597, as Richard Hooker, *Of the lavves of ecclesiasticall politie. The fift booke* (London, 1597; STC 13712.5); the passage appears in that edition on sig. A3r.

The reference to 'a small spark' as a potentially great danger to a natural body which might be 'overthrown through diseases bred within themselves' suggests that Harriot was, among other things, considering the life-threatening cancerous ulcer in his nose – the '*nihilum* [that] killed him at last', as someone cruelly joked when he died. He had begun treatment for it in 1615.[109]

Hooker had used his analogy with natural bodies to comment on diseases in bodies politic, particularly religious 'controversies'. He recognized two kinds. Those with the 'Church of Rome' represented external threats. While dangerous, he said they cause 'forces at home to be more united' and encourage 'virtuous minds' to hold 'contrary dispositions in suspense'. The dangers from 'domesticall euils' were greater, because people believe that they can 'readily master' them; they therefore permit them 'to run on forward till it be too late to recall them'. Hooker had in mind the Presbyterians or Puritans and religious radicals he associated with Thomas Cartwright.[110] By Harriot's day, the issues had shifted, as is suggested by many of the other books he acquired in 1617 along with Hooker's *Lawes*. Several are related to religious controversies between the Remonstrants and Counter-Remonstrants in the Netherlands; two came from the pen of Marco Antonio De Dominis, the Venetian ecumenist and sometime Anglican; two focused on Pelagianism with which the Remonstrants were tarred; others came from the anti-Calvinist party in the Church of England, including Lancelot Andrewes, and one from the anti-Presbyterian party in Scotland.[111] The religious issues raised in these works had come sharply into focus in 1617 with mob attacks on the Remonstrant churches by the Counter-Remonstrants in the Netherlands.[112] We can also see that Harriot was following the fallout very closely from the fact that he possessed a copy of a document in the hand of the anti-Calvinist Bishop John Overall or his secretary concerning theological issues in debate in the Netherlands at the time of the Synod of Dort (1618–19).[113]

---

109 John Aubrey recounts this jibe connecting Harriot's cancerous ulcer with what were thought to be his beliefs about Creation. It was passed to him by Seth Ward long after Harriot's death; see Aubrey, *Brief lives*, vol. 1, 110. Harriot was treated for the ulcer by Dr Theodore de Mayerne, starting in May 1615; see the drafts of his letters to Mayerne, dated 4 November 1615 and 5 April 1616, BL Add. MS 6789, ff. 446r-v, 447r.

110 Hooker, *Lawes*, Book 5, sig. V2v-[4]r.

111 PHA, HMC 241/4 f. 9r-v; Mandelbrote, 'Religion of Thomas Harriot', 253–54, 272–79.

112 See Jonathan Israel, *The Dutch Republic. Its rise, greatness, and fall, 1477–1806* (Oxford, 1995), 421–77.

113 BL Add. MS 6789, f. 464r: '*Tumque sunt Articuli in Belgio contraversi*'. Cambridge University Library (CUL), MS Gg/1/29 ff. 6r-7v is a lengthier version in the same hand. Alternative truncated versions, also in the same hand, can be found in CUL, MS Gg/1/7 v and Oxford, Bodleian Library, Tanner MS 279; see John Overall, 'On the Five Articles disputed in the Low Countries', in *The British delegation and the Synod of Dort (1618-1619)*, ed. Anthony Milton (Church of England Record Society, 13, 2005), 64–70.

Psalm 116/117 and Aquinas's poem, to which Harriot refers, address the need for religious peace. *Laudate Dominum* asks for unity: the Psalmist calls upon *all* nations and *all* peoples to join together in praise of God for His merciful kindness and enduring truth. Harriot's substitute, *Lauda Sion Salvatorem*, looks to Christ as the spiritual physician needed to cure the world of its doubt and heal it back to wholeness. It sings of Christ's body and blood hidden in the consecrated Host beneath the signs that are all that our eyes can actually see. The truth lies hidden from sight but is certain for Christians. In an allusion to the relationship between the Host and the *corpus mysticum* of the Christian commonwealth, *Lauda Sion* then warns against harboring doubt about the persistence of the Host's inherent unity even after it is broken into separate parts. Given the symbolic relationship between the consecrated Host and the *corpus mysticum* of society and the church, the invocation of the hymn speaks directly to the disunity into which the Christian community had fallen in Harriot's day and the need to heal it to wholeness and peace.[114] Aquinas ends the poem by calling on Jesus to strengthen us against death and grant that we may one day see Him face to face.[115] The final line alludes, of course, to 1 Corinthians 13:12: 'For now we see through a glass, darkly but then face to face: now I know in part; but then shall I know even as also I am known'.

We can now return to what initiated his train of thought – the first words of *Genesis*. In the course of Harriot's work, he experimented in several places in linking together words and letters written in his phonetic alphabet with those opening words about Creation. One, undated, displays Harriot's own name written twice in his phonetic alphabet, once all in lower-case letters and once with the initial letters as capitals – '*tomas haryots*' and '*Tomas Haryots*' – along with 'in the beginning' in the same phonetic alphabet. Another sheet, also undated, shows a string of individual phonetic letters, some words in Hebrew, including the first words of *Genesis* (some with and some without the vowel points), plus the Tetragrammaton written twice. In addition, there is a single sentence in the phonetic alphabet. Transliterated, it reads: '*ol thengz that have beeng hav caz wherefor they ar as they ar*' – i.e., 'All things that have being have cause wherefore they are as they are'.[116] On the reverse side of the same sheet, Harriot again wrote down the first words from *Genesis* 1:1 in his phonetic alphabet.[117] The two ideas converge: God's Creation from nothing results in the fixed certainty of all created things.

---

114 In Corpus Christi celebrations, the Christian community was represented as a unified whole composed of mutually supportive parts; see Mervyn James, 'Ritual, drama, and social body in the late medieval town', *Past & present*, no. 98 (February 1983), 3–29; Miri James, *Corpus Christi. The Eucharist in late medieval culture* (Cambridge, 1991).
115 equentia: 'Lauda Sion saluatorum', Missale Romanum (1577), lii.
116 BL Add. MS 6789, f. 494r (undated).
117 BL Add. MS 6789, f. 494 v (undated).

Harriot's most systematic expression and explanation of his phonetic alphabet appears in table form in a document presently in the possession of Westminster School. It bears the heading: 'An vniversall Alphabet conteyninge six & thirty letters, whereby may be expressed the lively images of Mans voyce in what language soeuer, first devised vpon occasion to seek for fit letters to expresse the Virginian speche, 1585'. At the top right a sequence of words, written in the phonetic alphabet, represents a translation of the Lord's Prayer as in the *Book of Common Prayer*: 'Our father which are in heav'n: hallow'd be thy name: thy Kingdom come'. Implicit are its remaining words as they appear in the *Prayer Book*: 'Thy will be done in earth, as in heaven. Give us this day our daily bread. And forgive us our trespasses, as we forgive them that trespass against us. And lead us not into temptation. But deliver us from evil. Amen'. The bottom of the table is then signed with Harriot's name, using the same phonetic alphabet.[118]

Anthony à Wood says that Harriot 'had strange thoughts of the scripture, and always undervalued the old story of creation of the world, and could never believe that trite position, *Ex nihilo nihil fit*'.[119] Creation *ex nihilo* entails that the universe is finite. For Epicurean atomists following Lucretius, such as Giordano Bruno, for whom atoms were eternal, the creation story in *Genesis* was a fiction or falsehood.[120] Although Harriot was a convinced atomist, he accepted its truth. In a brief thought experiment, probably from later his life, he juxtaposed contradictory conceptions in considering its logic. On the left-hand side of a folio he wrote: 'Nothing comes out of nothing: nothing further comes out of one; a third thing cannot come out of two, nor from a third, and so on ... there is nothing without movement; there is no movement without mass; therefore nothing comes from nothing'.[121] On the right-hand side is the alternative: 'From nothing comes nothing, from the one, nothing other, from duality no third, from the

---

118 Westminster School Library and Archives, Thomas Harriot's 'Alphabet'. Harriot's 'table' is reproduced as Figure 1 in Jacqueline Stedall, 'Symbolism, combinations, and visual imagery in the mathematics of Thomas Harriot', *Historia mathematica*, 34 (2007), 382; for the provenance of this document, see Salmon, 'Thomas Harriot', 151. I am grateful to Elizabeth Wells, Westminster School Archivist, for access to the original document and advice on its provenance. For the transliteration of phonetic script at the upper right, see Ian James, 'Universall alphabet of Thomas Harriot', http://skyknowledge.com/harriot.htm (June 2012).

119 Wood, *Athenae Oxonienses*, vol. 1, 300–1.

120 See, for example, Giordano Bruno, *De l'infinito universe et mundi* (Venice [London]: John Charlewood, 1584; STC 3938); *idem*, *De innumerabilis, immenso et infigurabili seu de universo et mundis* (Frankurt-am-Main: apud Johann Wechel et Peter Fischer, 1591; USTC 668230); Hilary Gatti, *Giordano Bruno and Renaissance science* (Ithaca, NY, 1999), 99–142; Ingrid D. Rowland, *Giordano Bruno. Philosopher/heretic* (New York, 2008), 214–22.

121 BL Add. MS 6788, f. 493r (undated); Gatti, 'Natural philosophy of Thomas Harriot', 71.

Trinity anything whatever'.[122] The latter formulation parallels the Huguenot poet Guillaume Du Bartas's paradox in his epic on the Creation: 'nothing is made from nothing ... since God created the world'.[123] Here Harriot shares Ralegh's convictions on the theme of Creation *ex nihilo*.[124]

In addressing these issues, Harriot was engaging with matters of immense complexity. In regard to the infinite, a topic to which he gave extensive attention, he asked apparently unanswerable questions: 'whether a finite may be resolved into indivisibles?'; 'whether the infinite be generated of finites?'; 'whether the infinite be composed of finites?'; and so forth. About all this he later pasted on a comment: 'Much ado about nothing/ Great warres & no blowes/Who is the foole now?' (see Figure 1.2).[125] 'Much ado about nothing' refers, almost certainly, to Shakespeare's play. 'Great warres & no blowes' succinctly and accurately summarizes it.[126] Harriot, who witnessed actual battles in his Virginia days, appears to have concluded that the debate about mathematical and metaphysical nothingness was similar to the 'mery war betwixt' Beatrice and Benedict in Shakespeare's play, who 'neuer meet, but there's a skirmish of wit betweene them'.[127] As might be expected from a play in which the names of the principals relate to 'blessedness', the comedy concludes happily with Beatrice and Benedick halting their verbal battles and publicly confessing their love for each other.[128] Peace triumphs along with love. Harriot appears to have decided that the quest for the discovery of the world

122 BL Add. MS 6788f, 493r (undated): *nihilo nihil fit/uno nihil aliud/duobus nihil tertium/tribus quodlibet*; trans. in John Henry, 'Thomas Harriot and atomism: a reappraisal', *History of science*, 20:4 (December 1982), 273.

123 'Le Second Iovr', in Guillaume de Saluste Sieur Du Bartas, *La Sepmaine ou Creation du Monde, Kritischer Text der Genfer Ausgabe von 1581*, ed. Kurt Reichenberger (Tübingen, 1963), 30–61; *Sepmaine* first appeared in 1578; the ninth Earl of Northumberland's library held a copy of the 1588 Geneva edition in duodecimo: Guillaume de Saluste du Bartas, *La Sepmaine ou Creation du Monde* (Geneva: pour Jacques Chouet, 1588; USTC 14786), PHA 5377, f. 45 v.

124 Ralegh, *History of the world*, 'Preface', sigs. D2r–E2v; see Jean Jacquot, 'Thomas Harriot's reputation for impiety', *Notes and records of the Royal Society of London*, 9:2 (May 1952), 178–80; Robert Hugh Kargon, *Atomism in England from Hariot to Newton* (Oxford, 1966), 18–30; Henry, 'Thomas Harriot and atomism', 271–73.

125 'De infinitis', BL Add. MS 6785, f. 436r (undated); Gatti, *Renaissance drama*, 64–65. Harriot worked extensively on the topic of infinity and may have contemplated publishing a treatise or commentary on it in mathematics and as a philosophical concept; in addition to BL Add. MS 6785, f, 436r, see BL Add. MS 6782, ff. 362r–374v.

126 Cf. Arianrhod, *Thomas Harriot*, 198–99. Galileo regularly compared 'intellectual debates to warfare'; David Wootton, *Galileo. Watcher of the skies* (New Haven, 2010), 160.

127 William Shakespeare, *Much Ado About Nothing*, Act I, scene 1, ll. 45–47. The play's text opens with a contrast between a real war which has just ended and the give-and-take between the two principals, Beatrice and Benedick.

128 *Much Ado About Nothing*, Act V, scene 4, ll. 71–116. Beatrice's name derives from Latin meaning 'one who blesses'; Benedick from *benedictus*, meaning 'blessed'.

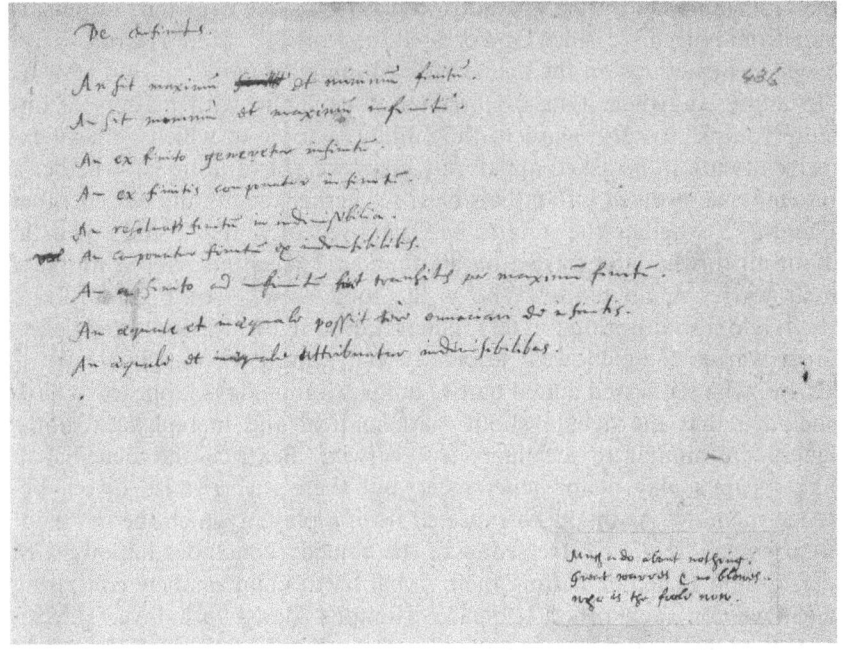

Figure 1.2 Harriot, undated. British Library, Add. MS 6785, f. 436r. By permission
of the British Library.

required a similarly irenical strategy – a love-thy-neighbor strategy – to
achieve its goal. 'Who is the foole now?' fits as well, since many of the
characters in Shakespeare's play call each other fool. But the source, more
probably, is the refrain of a drinking song, usually given the title 'Martin
said to his man', which was published in 1609 by Thomas Ravenscroft, a
person very likely known to Harriot.[129] The drunken ditty sings of *seeing*
unlikely things: 'a hare chase a hound', 'a mouse catch a rat', and, perhaps
most relevant to Harriot, 'a man in the moon': 'I see a man in the moon/
Clouting of Saint Peter's shoon ... Who's the foole now?'[130] 'Who is the
foole now?' is Harriot's self-reflective joke: he had spent many hours
looking at the moon through his telescope. Although he had not seen a

---

129 Thomas Ravenscroft (1588?–1635) was a boy chorister in St Paul's Cathedral and a
   member of the Children of Paul's, for which George Chapman, Harriot's friend, supplied
   material; see David Mateer, 'Ravenscroft, Thomas (*b.* 1591/2)', *Oxford dictionary of national biography*, Oxford University Press, 2004 [http://www.oxforddnb.com/view/article/
   23172].
130 Thomas Ravenscroft, *Deuteromelia. Or, The second part of the musicks melodie* (London,
   1609; STC 20757), no. 16.

'man in the moon' repairing St Peter's shoes, he could nonetheless draw the moon's features in some detail. Whatever the puzzles posed by the concept of infinity, he could still get on with his observations and see something definite (see Figure 1.3).

What then of *Genesis*? For an educated person in Harriot's day, Galileo for example, the starting place for examining it would have been the writings of St Augustine, who addressed the meaning of its description of Creation in

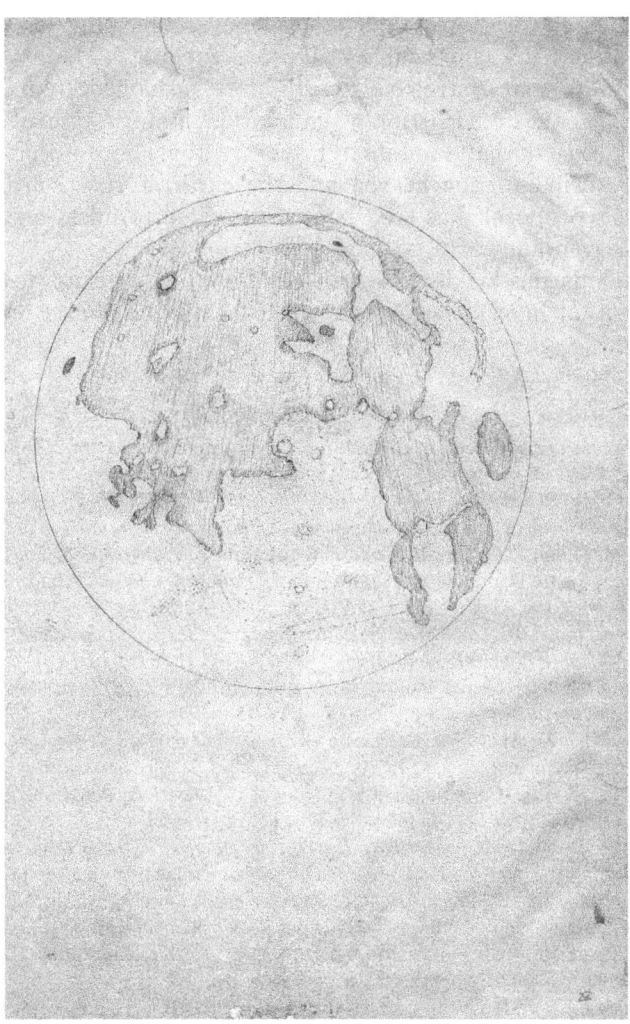

Figure 1.3 Harriot, moon map, c. 1610–11. Petworth House Archives, HMC 241 IX, f. 28r. By permission of Lord Egremont.

a number of his works.[131] Harriot himself was a close and careful reader of Augustine's writings.[132] For the interpretation of *Genesis*, the most accessible text was Augustine's *Confessions*, the final three books of which are devoted to examining and reflecting on its first words.[133] In those final books, Augustine repeatedly quotes *Genesis* 1:1 this way in his Latin text: *in principio fecisti caelum et terram*: 'in the beginning God *made* heaven and earth'.[134] As we have seen, Harriot similarly renders the passage in English utilizing the word 'made', instead of employing the verb 'created' as in all standard translations. The word in the Vulgate is '*creavit*'. English translations from the period – the Mathew Bible, the Great Bible, the Bishop's Bible, the Geneva Bible, the Douai-Rheims Bible, and the King James Bible – all use 'created'.[135] 'Made' leaves open the possibility that heaven and earth were fashioned from something already existent. For Harriot, a committed atomist, use of the word 'made' suggested that God first brought atoms into being and then fashioned them into the heaven and the earth, a view consistent with Augustine's proposal that God had first created formless matter before making heaven and earth.[136]

Augustine begins his discussion of *Genesis* with an acknowledgement of the 'poverty of human intelligence' when facing 'the impact made by the

131 John Hammond Taylor, 'Introduction: Augustine's studies on Genesis', in St Augustine, *The literal meaning of Genesis*, ed. and trans. John Hammond Taylor, 2 vols. (New York, 1982), vol. 1, 1–9. The works in which Augustine considered *Genesis* are: *Confessions*, esp. books 11–13; *Two Books on Genesis against the Manichees*; *The literal meaning of Genesis [An unfinished book]*; and *The literal meaning of Genesis. A commentary in twelve books*; see Ernan McMullin, 'Galileo's theological venture', in *The Church and Galileo*, ed. Ernan McMullin (Notre Dame, 2005), 88–116; John L. Heilbron, *Galileo* (Oxford, 2010), 200–23, esp. 211; Wootton, *Galileo*, 148.

132 See, for example, BL Add. MS 6789, f. 463r (undated); for Harriot's access to Augustine's works, see *supra* note 95.

133 Similar ground is covered in Book One of Augustine's *De Genesi ad litteram*, in St Augustine, *The literal meaning of Genesis*, 1:19–45.

134 See, for example, Augustine, *Confessions* XI.iii.5: *Audiam et intellegam, quomodo in principio fecisti caelum et terram*. In Augustine, *Confessions*, ed. and trans. Carolyn J. B. Hammond, 2 vols. (Cambridge, MA, 2016), vol. 2, 196–97; St Augustine regularly used forms of the verb "*facio*"—to make—for the act of creation.

135 The Vulgate reads '*in principio creavit Deus caelum et terram*'. The Nicean Creed prescribed in the Latin version of the *Book of Common Prayer*, used, e.g. in the Communion service, speaks of God as '*creatorem coeli & terre*'; *Liber precvm pvblicarvm*, sig, Kk3r. However, the Creed prescribed in the English text of the *Book of Common Prayer*, speaks of 'God yᵉ father' as 'almighty *maker* of heauen & earth'; *The booke of common praier* (London. 1559; STC 16292), sig. M2v (emphasis added).

136 *Confessions* XII.vii.7; Chadwick, 249. In *Confessions* XII.iii.3 and elsewhere Augustine speaks of the earth in *Genesis*, 1:2 as *invisibilis et incomposita*. As Augustine interpreted the passage, God first created formless matter, out of which he then made heaven and earth. The Geneva Bible's gloss reads: 'First of all, & before that anie creature was, God *made* heauen and earth of nothing' (emphasis added); Geneva version 1560, f. 1r.

words ... of holy scripture'.[137] He accepts, in particular, that disagreement could arise concerning the 'intention of the author' when 'using signs'. 'It is one thing', he says, 'to inquire into the truth about the origin of creation', which he regarded as beyond doubt, 'another to ask what understanding of the words ... was intended by Moses'.[138] A number of interpretations could be 'true statements', without one being able to see clearly 'which of them Moses had in mind', even though some commentators claim the opposite.[139] 'They have no knowledge of Moses opinion at all', Augustine says, 'but love their own opinion not because it is true, but because it is their own'.[140] When a difference of view arises, Augustine therefore advocated making a 'brotherly and conciliatory reply' based on Jesus's admonition to 'love the Lord our God ... and our neighbour as ourselves'.[141] Quarrelling 'about words', he believed, 'is good for nothing but the subversion of the hearers'.[142]

As we have seen, Harriot used his phonetic alphabet to place his own name on the sequence of thoughts triggered off for him by the first words of *Genesis,* suggesting it may have been added last to *seal* the document, as it were. Unlike the other items, this one breaks the pattern by using vowels. The squiggles signify the material certainty of the sounds made by the human voice when saying his name. They have a reality lacking in the ordinary English alphabet, since representing sounds, they remain the same, identifying the same person regardless of spelling, a point of special relevance in Harriot's day when, like the names of many of his contemporaries, his appears in a number of variant spellings, including in documents written in his own hand. Although sounds require listeners to be sounds, they have the same relationship to the words they represent as musical notation has to the sounds made by the human voice as, for example, when singing '*Salve Sion Salvatorem*' or 'Who's the foole now?'. There can be no quarreling about their existence, only about their meaning. Thomas Harriot, the person, remains Thomas Harriot, whether or not he is regarded as a learned mathematician, astronomer, and natural philosopher, or as Walter Ralegh's juggler or the Earl of Northumberland's magician. It is also of interest that elsewhere in his manuscripts Harriot spelled out the first words of *Genesis* in English using the same phonetic alphabet rather than Hebrew, again associating it in this instance with his own name, also written in the same phonetic alphabet.[143]

137 *Confessions* XII.i.1; Chadwick, 246; see also *Confessions* XII.xiii.16; Chadwick, 253.
138 *Confessions* XII.xxiii.32; Chadwick, 263.
139 *Confessions* XII.xxiv.33; Chadwick, 264.
140 *Confessions* XII.xxv.34; Chadwick. 264.
141 *Confessions* XII.xxv.35; Chadwick, 265; the scriptural citation is to Matthew, 22:37–9.
142 *Confessions* XII.xvii.27, Chadwick, 259; Augustine here cites 2 Tim. 2:14.
143 BL Add. MS 6787, f. 377 v; Harriot's name appears once as 'tomas haryots' and again as 'Tomas Haryots'.

In probing the meaning of *Genesis* 1 in light of his life-threatening illness and of the religious conflicts of his day, Harriot, in effect, was undertaking his own form of confession, meditating on a cure that would ease both his own condition and the condition into which the Church had fallen in his days. His model is summarized by Augustine's irenical position:

> [W]hen one person has said 'Moses thought what I say', and another 'No, what I say', I think it more religious in spirit to say 'Why not rather say both, if both are true?' And if anyone sees a third or fourth and a further truth in these words, why not believe that Moses discerned all these things? For through him the one God has tempered the sacred books in the interpretations of many who could come to see a diversity of truths.[144]

In these remarks St Augustine asks those struggling with theological puzzles to acknowledge limitations to their own understandings and to show goodwill to whoever sincerely sought the truth. A similarly irenical philosophical and religious outlook shaped Harriot's own search for discovery when he found it impossible to arrive at the certainty of a self-evident truth. 'Who is the foole now?' refers as much to himself as to others who also struggled with the dizzying puzzles posed by Creation.

The inscription on Harriot's funerary monument in London's Church of St Christopher le Stocks, speaks of him as *Veritatis indagator studiosissimus/ Dei Trini-unius cultor piissimus*: 'A most studious searcher after truth/A most devout worshipper of the Triune God'.[145] The testamentary clause of Harriot's last will also acknowledges his acceptance of the Trinity: 'I Comitte my Soule to the hands of Almighty God my maker and of his sonne Jesus Christe my Redeemer, of whose merritts by his grace wrought in mee by the holy Ghoste I doubte not but that I am made p*ar*taker, to thend that I may enioye the Kingdome of heaven pr*e*pared for the electe'.[146] However religiously skeptical Harriot might have been as a young man – claims that he was are perhaps overstated – by the last stage of his life, he was an entirely orthodox Christian.

144 *Confessions* XII.xxxi.42; Chadwick, 270–1.
145 The inscription on Harriot's funerary monument is printed in John Stow, *The survey of London*, ed. A[nthony] M[unday] and H[umphrey], D[yson] (London, 1633; STC 23345), 831-32, quoted at 832; Shirley, *Thomas Harriot*, 474. See *supra*, note 82.
146 The Will of Thomas Harriot, 29 June 1621; probated Archdeaconary Court of London, 6 July 1621; Guildhall Library, London MS 9053, box V (1618–1623) and MS 5051, vol. vi (Archdeaconary of London Court Records); CLR, Arch. Lond. 1618–1626/7; f. 71; in Stevens, *Thomas Harriot*, 194; a facsimile is printed in *A source book for the study of Thomas Harriot*, ed. J. W. Shirley (New York, 1981).

Conclusion: certain and full discovery

This essay has explored some of the ways in which Hakluyt and Harriot sought the 'certain and full discovery of the world', a subject whose pursuit they did not separate from their religious beliefs. For both, discoveries about the world using reason and the senses had the power to resolve profound controversies about God's works, discern truths about Creation, and provide a basis for tolerance, agreement, and the making of religious peace.

The foundations of Harriot's irenicism are apparent in his account of the peoples of Ralegh's Virginia, whose peoples he thought could be won by amity and persuasion to peaceable relations with the English and Christianity. Although he was not a Pelagian, he had faith that with goodwill the Virginians, and all non-believers, could be persuaded to accept the truths of Christianity as necessary and beneficial.[147] Although he spoke in his last will of joining the elect in heaven, he did not adhere strictly to the doctrine of double predestination.[148] As we have seen, he also drew freely on Roman Catholic traditions in his religious thought. Hakluyt similarly saw all humans as having the same desires for wellbeing and the same capacities to deliberate about the best means to achieve them, and therefore the same freedom to accept Christianity as their saving faith. Like Richard Hooker, he rejected the idea that the Church of Rome was a false church.[149] Hence he did not seek its extirpation, but joined William the Silent and his followers in the Netherlands, and later his patrons Sir Robert Cecil and Lord Charles Howard, the Lord Admiral, in believing that peaceable means could be found to accommodate Protestants and Catholics to one another.[150] For Harriot and Hakluyt, beneficial exchange – not just of commodities but of useful knowledge – would create the conditions of cooperation and mutual dependence needed to settle differences among Christians, win over non-Christians to belief, and pave the way for the restoration of the world to unity.

In the face of the deep religious controversies that inflamed their era and could not be brought to reconciliation by their 'plain aspect' or through 'invincible demonstration', the strategy of peace that Hakluyt and Harriot

---

147 Harriot, *Briefe and true report*, 25 and sig. D2r.

148 In regard to the testamentary clause in Harriot's will, Mandelbrote says this formulation reveals a 'Calvinist orthodoxy' in its 'opinions of the Trinity, the atonement, and the doctrine of election'; Mandelbrote, 'Religion of Thomas Harriot', 246. However, 'Calvinism' in Harriot's day was not an internally consistent, monolithic theology; there is nothing in his formulation that would have troubled Lancelot Andrewes.

149 See Sacks, 'Discourses of western planting', 447–48; Richard Bauckham, 'Hooker, Travers, and the Church of Rome', *Journal of ecclesiastical history*, 29:1 (January 1978), 42–50. Cf. Armitage, *Ideological origins*, 76–78.

150 See Sacks, 'Discourses of western planting', 428–30, 437–41, and the works cited there.

adopted was Erasmian in character.[151] From where might they have first been introduced to their Erasmianism? John Shirley offers a possibility in identifying Antonio del Corro as the figure assigned to catechize students in St Mary Hall while Harriot studied there.[152] Hakluyt also had a connection to Corro at Christ Church, where from 1578 to 1586 the Spaniard was *censor theologicae*, charged with overseeing the college's divinity students, Hakluyt among them; while there he also had dining rights as a 'commoner'.[153]

Corro's religious outlook was the product of the Erasmianism that dominated the monastery of San Isidore del Campo, where he began as a monk and which he left, along with a number of fellow monks in 1557, to avoid the attentions of the Spanish Inquisition. From the time he fled, his relations were contentious with the leaders of the Reformed Church in France and the Netherlands and with their allies among English Puritans, who suspected him of holding Pelagian views on salvation and perhaps of being an anti-trinitarian. Doubts, including about his moral character as well as his theology, had followed him to London. From 1571 to 1576, he was Divinity Reader in Temple Church, where his superior, the Master of the Temple, Richard Alvey, who had been a Marian exile, told Archbishop Parker that many who attended Corro's sermons regarded his views on free will and predestination as unorthodox; he associated himself with the same judgment. Those same doubts persisted when Corro then moved to Oxford in 1578 and was installed in Christ Church. Shirley suggests that Harriot's 'sensitivity' to revealing his 'religious scruples', and similar responses from others in Oxford who came into contact with Corro, arose from lessons learned from the 'ferment which surrounded his appointment' there, and that in Harriot's case at least, his story contributed with what Shirley argues was his 'sensitivity to the religious scruples of the time'. Otherwise, in Shirley's view, what he and his fellow students learned from Corro 'remains open to conjecture'.[154] We can say more. Soon after Corro arrived in London, and already under challenge from the leaders of the French Church there, he published two editions of his *Tableau de l'oeuure de Dieu* in an

151   See Hugh Trevor-Roper, "The religious origins of the Enlightenment', in *idem, The crisis of the seventeenth century. Religion, the Reformation and social change* (Indianapolis, 2001), 179–218; *idem*, 'The Ecumenical Movement and the English Church, 1560–1640', Wiles Lectures, May 1975; unpublished Dacre MSS, Christ Church, Oxford, File Soc.Dacre 2/5. Wiles Lectures 1975. I am grateful to the Dacre Estate and to Professor Blair Worden, Professor Trevor-Roper's literary executor, for permission to consult this manuscript.

152   Shirley, *Thomas Harriot*, 56–58.

153   Corro succeeded Lawrence Humphrey in the post. The first entry registering Corro's appointment is Christ Church Archives, *Christ Church Disbursements, 1578–1579*, Ch.Ch. MS xii.b.21, f. 22 v. Prior to joining Christ Church, Corro served as Divinity Reader in the Temple Church in London; *The Middle Temple Bench book*, ed. J. Bruce Williamson, 2nd edn. (London, 1937), 307.

154   Shirley, *Thomas Harriot*, 58.

effort to explain his religious views.[155] This text reveals him to have been an Erasmian universalist in his religious outlook. At the time, however, the text only lent credence to charges of Pelagianism against him. In the words of Patrick Collinson, 'he appears to posterity as a proto-Arminian', harbinger of the anti-Calvinist element in the Church of England whose books Harriot acquired in 1617 and who peopled Westminster Abbey's chapter while Hakluyt was a Canon there. As Collinson says, 'when Richard Hooker's views first became known ... comparisons were at once drawn with the Spaniard'.[156] Carro never fully escaped suspicions from Calvinists about the orthodoxy of his beliefs.

While at the Temple, Corro lectured first on St Paul's *Epistle to the Romans* in the hope of clearing the doubts raised about his beliefs by his *Tableau*. He then transformed the lectures into a dialogue,[157] which subsequently became the basis for his catechetical lectures to Oxford undergraduates, where Harriot would have heard them.[158] The publication of the former included as an appendix a set of 'Articles of the Catholike faith'.[159] Although it was not reprinted, the main text of the catechism incorporated

155 Antonio del Corro, *Tableau de l'oeuure de Dieu* (Norwich, 1569; STC 5792); Antonio del Corro, *Tableau de l'oeuure de Dieu* (London, 1570; STC 5792.5). The 1570 edition of the *Tableau* was dedicated '*A la tres-noble Dame, Madame de Stafford*', who had been a Marian exile in Geneva and Basel and then became Mistress of the Robes in Queen Elizabeth I's Privy Chamber; she was the mother of Sir Edward Stafford, with whom Hakluyt would serve as chaplain in Paris between 1583 and 1588.

156 Patrick Collinson, *Archbishop Grindal, 1519–1583. The struggle for a reformed church* (Berkeley and Los Angeles, CA, 1979), 146–51, esp. 151; N. R. N. Tyacke, 'Arminianism in England in religion and politics from 1604 to 1640', D. Phil. thesis (University of Oxford, 1969), 83–88; *idem, Anti-Calvinists. The rise of English Arminianism, c. 1590–1640* (Oxford, 1987), 58; Edward Boehmer, *Bibliotheca Wiffeniana. Spanish reformers of the two centuries from 1520. Their lives and writings, according to the late Benjamin B. Wiffen*, 3 vols. (Strasbourg, 1874–1904), vol. 3, 74. The comparison of Hooker to Corro was drawn by Walter Travers, who was Divinity Reader in the Temple from 1581 to 1586.

157 Antonio del Corro (Antonius Corranus), *Dialogus theologicus quo epistola Diui Pauli apostoli ad Romanos explanatur* (London, 1574; STC 5784); Antonio del Corro, *A theological dialogue Wherin the Epistle of S. Paul the Apostle to the Romanes is expounded* (London, 1575; STC 5786). As Roldan-Figueroa argues, Corro's views on God's universal redemptive grace derived especially from Andreas Osiander; see Rady Roldan-Figueroa, 'Antonio del Corro and Paul as the Apostle of the Gospel of Universal Redemption', in *A companion to Paul in the Reformation*, ed. R. Ward Holder (Leiden, 2009), 387–425.

158 Antonio del Corro to William James, Vice Chancellor of Oxford, 1 April 1581; Antonio del Corro, *Epistola beati Pauli apostoli ad Romanos,* (London, 1581; STC 5785), sig. a2r–a[5]v; see also Corro's preface '*Ivventvti Oxoniensis Academicae*', sig. a[6]r–b[3]v. Corro performed the service of *lector catechismi* in Hart Hall and Gloucester Hall as well as St Mary Hall, and also in St John's College.

159 Corro, *Dialogus theologicus*, ff. 96r–107r; Corro, *A theological dialogue*, ff. 139v–155v. Another edition of the Latin text of this *Dialogue* was published in Frankfurt in 1587: Antonio del Corro, *Dialogus in epistolam D. Pauli ad Romanos* (Frankfurt am Main: ex Officina typographica Nikolaus Basse, 1587; USTC 636123); this edition does not reproduce Corro's earlier appendix.

its principal doctrines. The 'Articles', then, represent a good starting place to discover what lessons Corro sought to offer to his Oxford students. On the question of Creation, his Latin text, taken without a citation from the version of Psalm 33:6 used by Heinrich Bullinger in the seventh chapter of *The Second Helvetic Confession*, said that the 'heavens were made' by 'the word of god': '*verbo dei coeli facti sunt*'.[160]

In the 'Articles', Corro, citing no contemporary theologian, made five substantive references to St Augustine in his interpretation of St Paul on sin, free will, and salvation.[161] In doing so, he took the view that Adam fell 'by hys owne fault' and, while he condemned the Pelagians, he did not completely reject the persistence of free will in human beings. Although he held that 'godly matters' require 'regeneration', i.e., 'the forgiuing of sinnes ... through God's 'grace', in 'outwarde things', as he put it, 'both the regenerated and the vnregenerated haue freewill'. In consequence, he argued that 'the vnderstanding of *earthly things* is not quite gone in man. For God of his mercy hath lefte him wit, howbeit farre vnlike that which was in him before his fall'. 'God', therefore, 'willeth men to manure their wittes, and addeth giftes and proceedings therwithal'.[162] Studies of the Book of Nature thus could yield truths with certainty. Corro also held 'that there is but one church ... Catholike and vniversall' from which 'no Nation ... is excluded', over which 'only Christ is the king'.[163] Along with rejecting the claims of the Papacy to supremacy over Christendom, this view is also a doctrine of peace. It regards the Church not just as universal, welcoming to all, but a commonwealth governed, implicitly, by Christ's two great commandments: to 'love the Lord God' and to 'love thy neighbour as thyself'.[164]

Corro's reasoning on this last point was modelled on Erasmus's *De sarcienda ecclesiae concordia*, 'On mending the peace of the Church', a text about how to restore all Christendom to peace and harmony, first published in 1533.[165] It is a lengthy commentary on Psalm 83/84, whose

160 Corro, *Epistola,* f. 98 v; Corro, *Theological dialogue,* f. 143r; Heinrich Bullinger, *Confessio et expositio simplex orthodoxae fidei, & dogmatu catholicorum syncerae religionis Christianae, concorditer ab ecclesiae Christi ministris, qui sunt in Helvetia, edita* (Zürich: excudebat Christoph Froschauer (II), 1566; USTC 624500), f. 7r.

161 For the references to St Augustine, see Corro, *Epistola,* ff. 101r (2), 103r, 103 v, 106 v; Corro, *Theological dialogue,* ff. 145 v, 146r, 148 v, 149r, 154r. Corro also made single, passing references to Origen, St Chrysostom, St Cyril, St Cyprian, and St Jerome.

162 Corro, *Epistola,* ff. 100r, 102r-v 103 v. 104r-v; Corro, *Theological dialogue,* ff. 144r, 147r-v, 148r, 149 v, 150r-v, 151 v (emphasis added).

163 Corro, *Epistola,* f. 98r; Corro, *Theological dialogue,* f. 142r.

164 Matthew, 22:37, 39 (KJV).

165 Desiderius Erasmus, *Liber de sarcienda ecclesiae concordia, dećque sedandis opinionum dissidiis, cum aliis nonnullis lectu dignis* (Basel: ex Officina Frobeniana,1533; USTC 635452); *idem,* 'On mending the peace of the Church', ed. and trans. Emily Kearns, in Desiderius Erasmus, *Expositions of the Psalms,* ed. Dominic Baker-Smith, 3 vols., vol. 3,

opening line in St Augustine's version, cited by Erasmus, reads: *Quam amabilia trabernacula tua, Domine virtutum*, 'How lovely (amiable) are your tabernacles'.[166] Erasmus's text builds its message on the Jewish celebration of Sukkot, the eight-day autumnal Feast of the Tabernacles. In Erasmus's hands, the Sukkah – the temporary dwelling housing the fruits of the harvest – becomes a figure for the Church.[167] Erasmus's central message is the prayer that

> God, who so governs human affairs in his inscrutable wisdom ... open the eyes of us all, so that seeing how lovely, how beautiful, how peaceful, secure and blessed are your tabernacles ... we may lay aside our differences of opinion and feeling, and, being in agreement in the same mind and the same opinion, we may so pass our days in the blessed society of all the saints that it may truly be said of us: 'Behold how good and how pleasant it is for brethren to dwell together in unity'.[168]

His sermon then concludes with a lengthy section outlining proposals for restoring the Church to peace with the aim not to claim for himself possession of 'certain truth, or to anticipate the decisions of the church', but to 'remove all causes of dissent' and not 'force on anyone a new religion which he finds abhorrent'.[169]

While in Antwerp in 1567, Corro published an *Epistre et amiable remonstrance* addressed to the Lutherans there in hopes of affecting their reconciliation with the adherents of the Reformed confessions in the city. His use of the adjective '*amiable*' is only one allusion to Augustine's version of the Psalm and to Erasmus's interpretation of it.[170] An English translation of this work was published in London in 1570, this time beginning with 'A prayer'. 'We crie with open throate', he says, 'that we are the Ambassadours of the holy Gospell of peace, and yet our attemptes and councels are ... but

---

Collected works of Erasmus, vol. 65 (Toronto, 2010), 125–216. This work's introduction in Latin and its running heads bear an alternative title: *de amabili ecclesiae Concordia*; Erasmus, *Liber de sarcienda ecclesiae concordia*, sig. a3r and ff.

166 The opening line of this Psalm in St Augustine's version, as cited by Erasmus, reads: *Quam amabilia trabernacula tua, Domine virtutum*. In the Vulgate, the word '*dilecta*', 'delightful', is used in place of '*amabilia*'.

167 The Jewish celebration of Sukkot is described in Leviticus 23:33–36 and in Numbers 29:12–38. Erasmus himself offers a brief description in 'On mending', *Collected works of Erasmus*, vol. 65, 137–38.

168 Erasmus, 'On mending', 65:197; the quoted passage is Psalm 132/33:1

169 Erasmus, 'On mending', 65: 213.

170 [Antonio del Corro], *Epistre et amiable remonstrance d'vn Minister de Euangile de nostre Redempteur IESVS CHRIST* ([Antwerp], s. n., 1567; USTC 4045). After Corro arrived in London, an English translation was published: Antonio del Corro, *An Epistle or Godlie Admonition of a Learned Minister of the Gospel of our Sauiour Christ*, trans. Geoffrey Fenton (London, 1570; STC 5788). In what follows I have followed Fenton's translation.

of warres, murders, and effusions of mans bloud'. We have 'Sathan in our harts'. To counter this, Corro calls for a 'vnitie of wils in diuine things' as well as 'concorde of doctrine'. 'In the beginning', he said, God 'created man to [His] own image to the end that he and his posterity might be one with [Him], hauing one only opinion led according' to His 'holy worde and manifestation and one wil affected to embrace that which [God] demands'. However, 'the serpent enemie of all peace ... a louer of dissention and discord, taught oure first Fathers the lesson of infidelitie, distrust, presumption and arrogant curioistie'. In consequence, 'by our workes we discouer a certeyne hate to our neyghbors', the very antithesis of Christ's call to 'love thy neighbour as thyself'. In response, his hope was that God would 'bring to pass ... that the fruits of justice may be sown in peace', so that 'by thy bountie thy wisedom from an highe ... may be pure, peaceable, moderate, tractable, full of mercie and good frutes, and far from debates, dissentions, and all hipocrisie'.[171]

Corro was himself a personal witness to the bloody consequences of religious war. He was on the walls of the city of Antwerp in March 1567 when government troops surprised and overwhelmed the 3,000 armed Calvinists encamped at Oosterweel, just outside the city. The result was a massacre, as Corro attests, when a number of Protestants, escaping through the fields to seek sanctuary, found Antwerp's gates locked against them, and 'brutish butcherie' occurred under Antwerp's wall; Corro described it in gruesome detail.[172] Hakluyt and Harriot also had personal experience of the realities and consequences of the religious upheaval of their time as well as of warfare. Both men knew intimately the truths embodied in Erasmus's *Complaint of peace* and both saw what transpired when peace was broken in the pursuit of transcendental goals. While Hakluyt was in service in Paris between 1583 and 1588, 'a dangerous tyme' as his supporters on the Privy Council said, he bore witness to the 'Thyestean tragedies' of religious war, including events in Paris preceding and following its 'day of the barricades' in 1588, which had driven the French king from the city.[173] Harriot knew the threat of violent death at first hand from his service in Ralegh's 'Virginia' colony in 1585–86, when the region's indigenous inhabitants placed him and his companions under attack; they escaped only when Sir Francis Drake, returning from his

---

171 'A prayer of the Author, for the concorde of doctrine, and vnitie in wils in diuine things, apperteyning to the aduauncement of the Gospell of CHRIST', Corro, *Epistle,*, sigs. *.3 v, *.[4]r-v, *.[5]r-v.

172 Antonio del Corro, *Lettre enuoyée a la Maiesté dv Roy des Espaignes* ([Antwerp]: s.n., 1567; USTC 10277), sigs. L2v-L4r; Corro's *Lettre* to Philip II was published in translation in London in 1577 together with another copy of the *Epistle*; Antonio del Corro, *A supplication exhibited to the most mightie Prince Philip King of Spain* (London, 1577; STC 5791).

173 Hakluyt, Epistle dedicatory to Sir Walter Ralegh', d'Anghiera, *De orbe novo*, sig. Hakluyt, sig. ā[5]v; Lords of the Council to Archbishop Whitgift, 18 May 1600, *APC*, 30:330–1;

West Indies raid, rescued them.[174] Harriot's experiences included not only association with the suspicions and troubles of Ralegh and the ninth Earl of Northumberland, before and after they were incarcerated in the Tower, but also, as we have seen, his own incarceration and interrogation for possible involvement in the Gunpowder Plot. Striving for peace was then personal for both the subjects of this essay.

Seeking the certain and full discovery of the world offered a way out of bloody conflicts of the kind Erasmus and Corro sought to end. Discovery would produce truths about the world as God had made it with a certainty requiring no other evidence than what was seen either with the eyes or could be proven logically. It would assert itself regardless of who offered it. The map of an island or seacoast drawn by a Catholic had the same truth value as one drawn by a Protestant. An astronomical observation or alchemical experiment was the same no matter the observer or experimenter, as was the proof of a mathematical theorem or the solution of an equation. Trust in the veracity of the reporter played a role. But it was about the veracity of the reported facts, including facts about the logical coherence of proofs, that others could observe and confirm or refute.

It is an optimistic view. As the era of new discoveries proceeded, the schools of skepticism – the Socratic, the Academic, and especially the Pyrrhonian – gained increasing influence, driving a conceptual wedge between truth discovered by their 'plain aspect' and truths discovered by 'invincible demonstration'. David Hume would later emphasize this distinction as a bright line that separated 'matters of fact' from 'relations of ideas'. The former are provisional rather than certain; they depend on 'the present testimony of our senses' and can be overturned by the evidence of new experiences.[175] Hakluyt and Harriot sought knowledge of facts that was as certain as knowledge of logic or mathematics. In doing so, their observations and researches paved the way to the world of knowledge-making in which we now live, and the contingent knowledge it produces.

174 Ralph Lane, 'An account of the particularities of the imployments of the English men left in Virginia by Sir Richard Greeneuill', Hakluyt, *PN* (1589), 737–47; Hakluyt, *PN* (1598–1600), vol. 3, 255–65; David Harris Sacks, 'Love and fear in the making of England's Atlantic empire', *Huntington Library quarterly,* 83:3 (September 2020), 543–66.
175 David Hume, *An enquiry concerning human understanding. A critical edition,* ed. Tom L. Beauchamp (Oxford, 2000), 24–28.

2 Thomas Harriot in the Twenty-First Century: 25 Years of the Harriot Lecture

Stephen Clucas

A little over 40 years ago, in 1974, John W. Shirley published a collection of essays, *Thomas Harriot: Renaissance scientist*, which presented Harriot as a 'key figure at the time when the new science of logic, reason, mathematics, and experiment was coming into being'.[1] In a rather uncharitable essay-review of this volume in the journal *History of science* in 1975, entitled 'In search of Thomas Harriot', the late Tom Whiteside criticized its contributors for not doing sufficient justice to their subject. If anything, Whiteside's praise of Harriot was more fulsome than any of the contributors to that volume. For Whiteside, Harriot 'possessed a depth and variety of technical expertise which gives him good title to have been England's – Britain's – greatest mathematical scientist before Newton'.[2] Coming from Whiteside, who knew Newton's manuscripts better than anybody at that time, this was high praise indeed; it came from a man who was not customarily given to hyperbole. Whiteside's review acknowledged the scholarship of John Shirley and Johannes Lohne in the 1950s which, he said gave us a 'true appreciation of Harriot's expert skill and inventiveness in the physical sciences'.[3] However, he did feel that some very pressing questions had been neglected by Shirley and his contributors. 'Such questions', he says, 'as how near did Harriot come to creating a viable theory of... "local" motion and collision... are not even asked, let alone answered'.[4]

In search of Thomas Harriot

A great deal of work has been done on Harriot since 1974, and yet in many ways we are still 'in search of Thomas Harriot'. In this chapter I reflect briefly on the advances made in our understanding of Harriot over the past 40 years, and celebrate the 25 years of Harriot lectures at Oriel, which have

---

1 J. W. Shirley (ed.), *Thomas Harriot. Renaissance scientist* (Oxford, 1974), viii.
2 D. T. Whiteside, 'In search of Thomas Harriot', *History of science*, 13 (1975), 61–70 (61).
3 Whiteside, 'In search of Thomas Harriot', 62.
4 Whiteside, 'In search of Thomas Harriot', 65.

DOI: 10.4324/9781003096580-3

made such a signal contribution to those advances. By surveying some of the significant moments in Harriot studies in the past decades I hope to come to an understanding of what Harriot means for the twenty-first century, and to reflect on what remains to be done in the decades to come. Much has been achieved, but many questions remain to be asked.

There is no question that one of the key areas of Harriot's achievement which has been properly elucidated since Shirley's time is Harriot's algebra. In her book *A discourse concerning algebra: English algebra to 1683*, published in 2002, the late Jacqueline Stedall reassessed the contribution that Harriot had made to this subject. The great Oxford mathematician John Wallis held Harriot in very high esteem, but he has been criticized by later historians – who only knew Harriot's posthumously published writings – for partiality or chauvinism.[5] A careful examination of Harriot's papers, however, suggested a different story: '[A]mongst the disarray of Harriot's manuscript sheets', Stedall wrote, 'we have the scattered pages of an invaluable treatise, a treatise that extended the contemporary understanding of polynomial equations'.[6] Wallis had somehow divined from the published version of Harriot's algebra what Harriot had really been trying to say. Wallis, Stedall argued, 'saw the true magnitude of what Harriot had done'.[7]

In *The greate invention of algebra*, published in 2003 – using some hints from Cecily Tanner and Muriel Seltman – Stedall was able to reconstruct this un-published treatise on equations from 140 scattered sheets in the surviving manuscripts.[8] Using Nathaniel Torporley's manuscript treatises the *Congestor analyticus* (Torporley's incomplete attempt at reconstructing Harriot's trea-tise) and the *Corrector analyticus artis posthumae Thomae Harrioti*, which criticizes Harriot's posthumously published treatise on algebra edited by Walter Warner (the *Artis analyticae praxis*), Stedall was able to present the fullest possible account of Harriot's algebraic solutions of polynomial equations.[9] Stedall's volume was complemented by the first republication of the *Praxis* since 1632, in the shape of Robert Goulding and Muriel Seltman's translation, published by Springer in 2007.[10] Both of these publications showed Harriot's initial debt to the algebraic work of the French mathema-tician François Viète, but also the extent of his own original contributions.[11]

---

5 J. A. Stedall, *A discourse concerning algebra. English algebra to 1685* (Oxford, 2002), 88.

6 Stedall, *English algebra*, 96.

7 Stedall, *English algebra*, 123.

8 J. A. Stedall, *The greate invention of algebra. Thomas Harriot's treatise on equations* (Oxford, 2003).

9 Stedall, *English algebra*, 107–11, and Stedall, *Greate invention*, 22–26.

10 *Thomas Harriot's Artis analyticae praxis. An English translation with commentary*, ed. M. Seltman and R. Goulding (New York, 2007).

11 See Stedall, *Greate invention*, Appendix, 'Correlations between Harriot's manuscripts and the texts of Viète, Warner and Torporley', 291–99, and Seltman and Goulding, *Artis analyticae praxis*, 'Comparative table of equations solved', 263–69, and 'Commentary', 209–62, *passim*.

Viète's algebra was, as Stedall observed, 'the foundation of Harriot's',[12] and he had privileged access to some of it even before it was published, via his friend Torporley, who had acted as Viète's amanuensis in the mid-1590s.[13] However, not only did Harriot rewrite sections of Viète's *De numerosa potestatum ad exegesin resolutione* (1600) in his own notation, but he also 'explored for himself the theoretical underpinnings of Viète's method, and so developed his own treatment of the structure and solution of polynomial equations'.[14] Harriot made particular advances, Stedall argued, in the relationships between roots and coefficients, and while Viète analysed equations in terms of ratios, Harriot 'saw the possibility of writing polynomials as products of factors of lower degree'.[15] He also developed an innovative method for removing a cube term from a quartic equation which is not found in the work of any of Harriot's predecessors.[16] Harriot emerges from Stedall's study, as a much more creative algebraist than those who judged him solely on the basis of his printed work imagined.

As Stedall's painstaking reconstruction of the treatise on polynomial equations suggests, Harriot's surviving manuscripts are thin in terms of freestanding manuscript treatises. However, Stedall – together with her colleague Janet Beery – was instrumental in publishing one such work, the *De numeris triangularibus. et inde de progressionibus arithmeticis magisteria magna*, which survives in British Library, Additional MS 6782. The *De numeris* is written almost entirely in notation, so in this project Stedall and Beery published the manuscript in facsimile (which is perfectly legible and has the added benefit of preserving the layout of the original), together with a long introductory essay.[17] The *De numeris* presents an algebraic method for interpolating tables by means of constant differences, which (as Stedall and Beery show) was shared and discussed by English mathematicians throughout the seventeenth century, although the means by which it was disseminated was largely informal.[18] Harriot found his way to his own arithmetical theories via those of Boethius and Jordanus (republished together in 1503 by Jacques Lefèvre d'Étaples) and those of more recent mathematicians such as Francesco Maurolico, Michael Stifel, and Girolamo Cardano.[19] Through these earlier mathematicians, Harriot became interested in the properties of triangular numbers, and specifically how they could be used to devise a method for

---

12 Stedall, *Greate invention*, 5.
13 Ibid., 4–5, and fn. 6, 301 for Stedall's dating of Torporley's letter to Harriot at the time he met Viète.
14 Ibid., p. 6. Cf. Stedall, *English algebra*, 94.
15 Stedall, *Greate invention*, 14.
16 Stedall, *Greate invention*, 17; Stedall, *English algebra*, 96.
17 *Thomas Harriot's doctrine of triangular numbers. The 'Magisteria magna'*, ed. J. Beery and J. A. Stedall, Heritage of European Mathematics (Zurich, 2009).
18 Beery and Stedall, *Triangular numbers*, 3–4.
19 Beery and Stedall, *Triangular numbers*, 5.

interpolation. That is to say, how given a difference table of arithmetic progressions, one could devise a method for interpolating new values between the existing values.[20] This method had many potential applications, and Stedall and Beery believe that Harriot began thinking about this problem during the period when he was working on tables of meridional parts (required to calculate the accurate position of a ship at sea following a constant compass bearing), and had arrived at an algebraic solution by 1614.[21] Harriot used his new method to generate polyhedral numbers,[22] and in his work on polynomial equations involving the sums of squares, cubes, or higher powers,[23] which were later to become important to the development of the integral calculus.[24] Stedall and Beery trace the dissemination of Harriot's method from his immediate colleagues Nathaniel Torperley and Walter Warner, to a wider circle of mathematicians, including Henry Briggs, Charles Cavendish, John Pell, John Collins, and Nicolaus Mercator.[25] Although it was not widely known, the fact that 'Harriot's method of differences was still in use amongst English mathematicians almost 60 years after Harriot invented it', is not without significance, and Harriot's method was not surpassed until the work of Newton and Gregory in the late seventeenth century.[26] These studies show that Harriot was a significant mathematician, whose work more than justified the high opinion entertained about it by Charles Cavendish and John Wallis.

Another heroic work of reconstruction, which has fulfilled some of the enthusiastic assessments of Shirley and his contemporaries, is Matthias Schemmel's *English Galileo*, published in 2008.[27] Like Stedall, Schemmel had literally to piece together all of the existing papers on free fall and ballistic trajectories, scattered throughout Harriot's papers – in this case running to over 180 folios. In doing so, Schemmel reveals Harriot to have been a significant figure in pre-Classical mechanics, someone who wrestled (like Galileo) with intractable problems using the mathematical and natural philosophical tools available to him at the time, and who used experiments and the 'shared knowledge' of practical men to shape his investigations.

Although – as has often been noted – most of Harriot's work remains unpublished, and could therefore be seen as a 'dead end in the history of science', Schemmel argues that Harriot's manuscripts gain in value when viewed from the perspective of 'historical epistemology'.[28] This approach – associated with

---

20 Beery and Stedall, *Triangular numbers*, 9.
21 Beery and Stedall, *Triangular numbers*, 15, 17.
22 Beery and Stedall, *Triangular numbers*, 17.
23 Beery and Stedall, *Triangular numbers*, 19.
24 Beery and Stedall, *Triangular numbers*, 20.
25 Beery and Stedall, *Triangular numbers*, 20–47.
26 Beery and Stedall, *Triangular numbers*, 47–52.
27 M. Schemmel, *The English Galileo. Thomas Harriot's work on motion as an example of preclassical mechanics*, 2 vols. (Dordrecht, 2008).
28 Schemmel, *English Galileo*, vol. 1, 4–5, 232.

the work of historians of science at the Max Planck Institute for the History of Science in Berlin – is less concerned with individual contributions to scientific knowledge, and focuses instead on the 'shared knowledge' of mathematical practitioners in early modern Europe, and how it facilitated the shift from pre-classical to classical mechanics.[29] From this viewpoint, 'major' figures like Galileo retain their significance, but the work of 'lesser known contemporaries' (like Harriot) are seen as important for establishing a fuller understanding of the historical process. While Schemmel does not want to argue that the English mathematician was a 'neglected Galileo', Harriot nonetheless emerges from this study with some credit. Harriot and Galileo are seen as similar figures, 'occupied with similar problems', both of whom have 'shortcomings when viewed from within the classical framework'.[30] Schemmel's close analysis of Harriot's research methods reveal a figure in whose work 'we [...can] discern several of the crucial insights for which Galileo was famous'.[31] Schemmel, however, seeks to avoid the anachronistic teleology of 'progress' towards classical mechanics, and the invidious canonical logic of comparing 'major' to 'minor' figures, attempting instead to show that both the English and the Italian mathematician drew on a shared legacy composed of Aristotelian physics, the mediaeval *calculatores*, ancient mathematics, and the knowledge of practitioners, such as engineers and gunners.[32] This shared knowledge, as he so aptly puts it, 'defined a space of possible alternative solutions' available to early modern thinkers.[33]

Schemmel shows how Harriot (like Galileo, Descartes, and Beeckman) approached the question of motion using the tools provided by the medieval *calculatores* and the geometrical representations of motion popularized by Nicolas Oresme,[34] and was able to arrive at a satisfactory understanding of the time-squared law, although he was ignorant of the 'conditions of its mathematical validity'.[35] Harriot, however, quickly realized that while these representations were adequate for capturing 'uniformly difform' motion with respect to time, they were unable to do the same with respect to space.[36] Searching amongst the known curves of the time, Harriot alighted (possibly with the aid of some hints from Thomas Digges's *Stratioticos* of 1579) upon the parabola as a likely candidate for representing projectile trajectories.[37]

29 Schemmel, *English Galileo*, vol. 1, 5.
30 Schemmel, *English Galileo*, vol. 1, 3–4.
31 Schemmel, *English Galileo*, vol. 1, 233.
32 Schemmel, *English Galileo*, vol. 1, 5.
33 Schemmel, *English Galileo*, vol. 1, 235.
34 Schemmel, *English Galileo*, vol. 1, 56.
35 Schemmel, *English Galileo*, vol. 1, 66.
36 Schemmel, *English Galileo*, vol. 1, 85–87.
37 Schemmel, *English Galileo*, vol. 1, 87. For Digges's speculations about ballistic trajectories and geometrical curves, see Thomas Digges, *An arithmeticall military treatise, named Stratioticos* (London, 1579), 'Of randons', 186–89.

Schemmel shows how Harriot turned to experiments with falling bullets and a balance to decide whether the law of motion of falling bodies was with respect to time, or to space, and used geometrical diagrams – which represented both his theoretical calculations (using arithmetical proportions and algebraic tools) and his experimental findings – as exploratory research tools rather than as mere descriptions of motions. By comparing calculated and experimental values, he was able to conclude that the motion of fall obeys the law of time proportionality.[38]

Schemmel also shows that, like Galileo, Harriot's work was enmeshed in 'practitioners' knowledge', and that it embodied the fruitful 'union of practical mathematician and natural philosopher' which was so important in this period.[39] Schemmel shows how Harriot began his research on projectile motion by harvesting the empirical findings from practical treatises by Niccolò Tartaglia, William Bourne, Luys Collado, and Alessandro Capobianco, rescaling their results in order to facilitate comparison.[40] He also shows that Harriot was prepared to rethink his mathematical results if they failed to capture practitioners' knowledge,[41] arriving at results that are 'amazingly similar' to modern trajectories.[42] Attentive as he was to practical knowledge, Schemmel's analysis reveals how Harriot made creative use of recently revived ancient mathematics in his work. Thus he used the theory of extrusion from Archimedes's *On floating bodies* to explain differences in the velocities of falling bodies in terms of buoyancy,[43] and used Apollonius's work on conic sections to establish maximum ranges in relation to angles of elevation,[44] and to prove that all projectile trajectories are parabolic.[45]

Schemmel's study thus shows how Harriot used the existing knowledge base available to him to move from a concept of motion which was basically Aristotelian in the 1590s to one that saw the compound motions of projectiles as a combination of the inclined plane and the balance by 1606, a move which effectively abolished the Aristotelian distinction between natural and violent motion.[46]

The reconstruction work of Stedall and Schemmel in 2003 and 2008 are probably the two most significant steps towards a fuller understanding of Harriot's legacy, and provide an inspiring example for those who follow after them. If this were not enough, Stedall and Schemmel, together with Robert Goulding, have acted as editors of the Harriot Online Project,

38 Schemmel, *English Galileo*, vol. 1, 113.
39 Schemmel, *English Galileo*, vol. 1, 19.
40 Schemmel, *English Galileo*, vol. 1, 27–29.
41 Schemmel, English Galileo, vol. 1, 161–64.
42 Schemmel, *English Galileo*, vol. 1, 175.
43 Schemmel, *English Galileo*, vol. 1, 136–41.
44 Schemmel, *English Galileo*, vol. 1, 187.
45 Schemmel, *English Galileo*, vol. 1, 207.
46 Schemmel, *English Galileo*, vol. 1, 225.

supported by a team of technical advisors at the Max Planck Institute, and currently hosted by the ECHO Cultural Heritage Online website. This project – which is still ongoing – seeks to make available high-quality digitized images of all of Harriot's papers, together with transcriptions and commentaries. As a tool for future Harriot researchers, the finished project will be of inestimable value, and make possible the kinds of reconstructive scholarship at which Stedall and Schemmel have excelled.

The Harriot lectures, 1990–2015

It is no surprise, then, to note that both Schemmel and Stedall made important contributions to the series of 23 lectures hosted by Oriel College, beginning with David Quinn's 'Thomas Harriot and the problem of America', delivered on 7 May 1990. A significant number of these lectures have now been published in two important volumes edited by the *éminence grise* of the Harriot Lecture, Professor Robert Fox: *Thomas Harriot: an Elizabethan man of science* published in 2000, and *Thomas Harriot and his world: mathematics, exploration, and natural philosophy in Early Modern England*, published in 2012.[47] The second of these two volumes includes pieces by Stedall and Schemmel on their larger projects, based on lectures they gave in 2002 and 2005 respectively. But the two volumes together present a wide range of other valuable perspectives on Harriot, which reflect recent historiographical shifts in the history of science and mathematics.

Obviously, given the constraints of space, I am unable to rehearse the themes of all 23 lectures, so what follows will of necessity be a brief, and rather partial response to those lectures which I feel have advanced our understanding of Harriot. I would like to begin, perversely, with one of the lectures that was not published. In 2007 Stephen Johnston's lecture 'Thomas Harriot and the English experience of navigation', drew our attention to an aspect of Harriot that has sometimes been forgotten: Harriot as a mathematical practitioner. Johnston situated Harriot in a world that his own work has done much to reconstruct. In his 1994 Ph.D. thesis 'Making mathematical practice',[48] Johnston radically overhauled Eva Taylor's concept of the 'mathematical practitioner',[49] seeking to identify a culture of mathematical practice in Elizabethan England, in which men from very different backgrounds engaged in a broad range of Renaissance mathematical activities, including

47 R. Fox (ed.) *Thomas Harriot. An Elizabethan man of science* (Aldershot and Burlington, VT, 2000); idem, *Thomas Harriot and his world. Mathematics, exploration, and natural philosophy in early modern England* (Farnham and Burlington, VT, 2012).

48 S. Johnston, 'Making mathematical practice: gentlemen, practitioners and artisans in Elizabethan England', Ph.D. thesis (University of Cambridge, 1994).

49 E. G. R. Taylor, *The mathematical practitioners of Tudor and Stuart England* (Cambridge, 1967).

surveying, navigation, gunnery, architecture, and shipbuilding.[50] Johnston's lecture made us think again about the impact that the Roanoke voyage and English maritime exploration more generally had on the young Oxford-trained mathematician. This aspect of Harriot was also stressed by Pascal Brioist in his 2009 Harriot lecture, 'Thomas Harriot and the worlds of practice', which looked in detail at Harriot's intense scrutinizing of the Elizabethan sailing ship as a complex assemblage of mechanical devices. Harriot's notes on sails and rigging, bear vivid testimony to what Brioist calls his 'special capacity to absorb all sorts of practical knowledge'.[51] The 1995 lecturer Jim Bennett, in his lecture on 'Thomas Harriot's place on the map of learning', brought Harriot into a complex but fruitful historiographical focus. The disappointment that some historians of science felt when confronted with Harriot's work, he argued, was largely the product of 'unhelpful perspectives' and 'inappropriate historiography'. Harriot, he said, 'had become a disputed property, caught between different ways of viewing the mathematics and natural philosophy of the period'.[52] While Whiteside unequivocally saw Harriot as a 'mathematical scientist', Bennett rejected such anachronism. 'Harriot the scientist is an impossible characterization for the period', he argued, whereas 'Harriot the mathematician' was 'clear and evident'.[53] Like Johnston, Bennett saw Harriot as a product of the 'remarkable vigour of practical mathematics... in the fifteenth and sixteenth centuries'. While he resisted the attempts of historians like Robert Kargon to piece together Harriot as a natural philosopher,[54] he was interested in what happens when Harriot the mathematician is drawn into natural philosophical questions.[55]

For me, one of the most enduring products of the Harriot lectures is precisely the exploration of this historiographical crux about Harriot as mathematician or natural philosopher. This became clear in the two lectures of 2004 and 2005. In 2004 Matthias Schemmel gave his lecture 'The English Galileo: Thomas Harriot and the force of shared knowledge in early modern mechanics', while in 2005, John Henry responded with a lecture entitled 'Why Thomas Harriot was not the English Galileo'. Schemmel and Henry have very different conceptions of Harriot's career. For Schemmel, Harriot's work on projectile trajectories and free fall makes an instructive parallel with the natural philosophical achievements of Galileo. Schemmel's historical epistemological approach attends to the different 'inferential pathways' followed by Harriot and Galileo in their works on mechanics based on

50 A similar approach is taken by Eric H. Ash in his *Power, knowledge, and expertise in Elizabethan England* (Baltimore, MD, 2004).
51 Fox (ed.), *Thomas Harriot and his world*, 200.
52 Fox (ed.), *Elizabethan man of science*, 139.
53 Ibid., 142.
54 Ibid., 139–41.
55 Ibid., 142.

a very similar body of 'shared knowledge'.[56] Working independently on the same 'challenging problems' in ballistics, Schemmel argues, Harriot and Galileo raised 'virtually identical questions' about central problems in mechanics.[57] Schemmel points out that Harriot's knowledge of free fall and the shape of ballistic trajectories by 1621 was essentially the same as Galileo's in 1638, with both employing the same inclined plane conception of projectile motion.[58] If anything, he argued, Harriot was 'more successful than Galileo in consistently relating the concept of velocity to the graphical representation of motion'.[59]

Henry does not explicitly refute Schemmel's claims. But he argues that while Galileo amalgamated 'speculative natural philosophy with mathematical and experimental traditions', Harriot 'remained first and foremost a mathematical practitioner'.[60] Though conceding that there are some elements of Harriot's work which could be construed as natural philosophical – his work on optics, impacts, and atomism – he says that 'Harriot was willing to play the natural philosopher sometimes'[61] and ultimately sees Harriot's forays into natural philosophy as either a failure or a symptom of his 'uncompromising perfectionism'.[62] For Henry, Harriot must be excluded from a list of seventeenth-century figures (like Galileo and Descartes) who were both mathematicians and natural philosophers, who mediated between what Jim Bennett has called 'the mechanics' philosophy' and the mechanical philosophy because his work shows little interest in causal explanations.[63]

Robert Goulding's 2001 lecture painted a very different picture. Harriot's work on optics, Goulding argues, 'were connected with his abiding natural philosophical concern with the structure of matter',[64] and he shows that Harriot's understanding of refraction was based on an atomistic model,[65] and that his optical experiments should be seen as a 'species of alchemical experimentation'.[66] The historical assessment of the sixteenth and seventeenth centuries, and the complex relationship between mathematics and natural philosophy, let alone the so-called scientific revolution, is still far from settled. Harriot, who was both a gifted mathematician and a scrupulous experimenter, is a figure who will help us to continue thinking through these important questions.

56 Fox (ed.), *Thomas Harriot and his world*, 106–10.
57 Ibid, 92.
58 Ibid, 98–101.
59 Ibid, 110.
60 Ibid., 117–18.
61 Ibid, 130.
62 Ibid, 134, 136–37.
63 Ibid., 132.
64 Ibid., 32.
65 Ibid. 41–43.
66 Ibid., 43.

Future prospects

So what remains to be done? Which areas of Harriot's legacy have still to be fully explored? I'd like to begin this section of my talk with a snapshot of my own recent – as yet unpublished – work on a very particular aspect of Harriot's work, his mechanics. The benefits of this are twofold. Firstly, it will show that there are unexplored aspects of topics which have already been well covered. Secondly, it emphasizes the fruitfulness of approaching the work of Harriot via his friends and colleagues, in this case the mathematician and natural philosopher, Walter Warner.

Harriot's small treatise on the collision of round bodies, the *De reflexione corporum rotundorum* was prepared in 1619 for his patron Henry Percy, ninth Earl of Northumberland from some 'auntient... notes' on the topic.[67] We know from a letter accompanying this treatise, that Harriot shared his ideas with both Warner, and another friend and member of the Percy household, Robert Hues:

> Sir. When Mr Warner and Mr Hues were last at Syon, it happened that I was perfecting my auntient papers notes of the doctrin of reflections of bodies. Unto whom I imparted the magisteryes thereof, to the end to make your Lordship acquainted with them as occasion served.[68]

Harriot's writings on the collision of round bodies are well-known and have been closely studied by Jon Pepper, Martin Kalmar, and (most recently) by Russell Smith.[69] However, a group of hitherto unknown manuscripts on collisions and motion written by Walter Warner provides us with new evidence that provides a new context for Harriot's mechanics.[70]

In the Isham-Lamport papers in the Northamptonshire Record Office, a significant number of previously undiscovered Warner manuscripts were unearthed in the mid-1990s by Timothy J. Raylor: some 337 folios in 14 sewn

67 Petworth House Archive (hereafter PHA), HMC MS 241 VIa, ff. 23r–31r.
68 Thomas Harriot to Henry Percy, 13 June 1619, London, British Library (hereafter BL), Harleian MS 6002, f. 21r–v. This is a copy made by Charles Cavendish, who adds the following note: 'Mr Harriot's letter to my Lo: Northumberland: annexed to his treatise of Reflections: lent me to transcribe by Sir Th: Alesburie'.
69 J. V. Pepper, 'Harriot's manuscript on the theory of impacts', *Annals of science*, 33 (1976), 131–51; M. Kalmar, 'Thomas Harriot's *De reflexione corporum rotundorum*: an early solution to the problem of impact', *Archive for history of exact sciences*, 16 (1977), 201–30; R. Smith, 'Optical reflections and mechanical rebound: the shift from analogy to axiomatization in the seventeenth century. Part 1', *British journal for the history of science*, 41 (2008), 1–18, esp. 7–15. There is also a recent doctoral thesis which devotes part of a chapter to the *De reflexione*: S. J. Hyslop, 'The mathematics of collision and the collision of mathematics in the 17th century', Ph.D thesis (Indiana University, 2015); see chapter 2, 'First investigations: Harriot and Beeckman on collision', 10–51.
70 What follows summarises some of the findings of my forthcoming article: 'Thomas Harriot and Walter Warner on collisions: mechanics and natural philosophy in early seventeenth century England'.

notebooks or smaller manuscripts, distributed into six manuscript bundles. Amongst these are three sewn notebooks relating to what Warner refers to as 'kinetica' or the 'doctrina de motu'. These are:

> Notebook III: Fragments of a treatise on the collision of spherical and plane-sided bodies.
> Notebook IV: A treatise labelled 'De motu et quiete'
> Notebook V: A set of notes headed 'De corporum non resultantiam ... effectis ex mutua incidentia oriundis' ['Effects arising from the mutual incidence of non-rebounding bodies'], including a number of theorems, definitions, consectaries, and problems concerning the collision of bodies.

These 39 folios – which can be dated to around 1601–1603 – give us a new insight into the study of mechanics in the Northumberland circle.

Warner's papers suggest a hitherto unacknowledged source for a key concept in Harriot's work on collisions – *linea nutus*, the line of inclination, or line of force, which Martin Kalmar glosses as 'the tendency that a body has to move in a given direction with a given speed'.[71] Both Warner and Harriot use this term,[72] and they both consider collision in terms of the composite of the weights and motions of the colliding bodies, represented by lines which designate 'the forces or active powers of given bodies'.[73] In his papers on rebounding and non-rebounding bodies, Warner describes the motion of two bodies 'after the moment of impact' (*post incidentiae ictum*) and the 'new velocity' (*velocitas nova*) or 'secondary velocity' (*velocitas secundaria*) acquired by the moved body.[74] Harriot's two porisms also talk of the 'effect after the moment [of impact]' (*effectus post ictum*) and the 'nutus acquired from the moment [of impact]' (*nutu ex ictu acquisito*), or 'second motion' (*motus secundus*) which is 'a composite of two forces' (*ex duobus nutibus compositus est*).[75] While he recognized its importance, Kalmar was mistaken is in his assumption that 'This use of *nutus* is ... unique to [Harriot]'.[76] As Jon Pepper has correctly noted, the *linea nutus* concept had a 'respectable

---

71 Kalmar, 'Problem of impact', 204. For a seventeenth-century definition, see M. Mersenne, *Universae geometriae, mixtaeque mathematicae synopsis, et bini refractionum demonstratarum tractatus* (Paris, 1644), II. 3, p. 458: 'Linea recta ab extremo semidiametro in diametrum perpendiculariter acta demonstrat qualis sit nutus; nam quo maior est semidiameter, eo maior est praedicata linea, praedictúsque nutus; linea enim recta ducta à termino, à quo mobile mouetur, vsque ad terminum, in quo quiescit, dicitur *linea nutus* [...]'.

72 It is also used by Nathaniel Torporley in his 'De pondere aquae', BL Add. MS 4458, ff. 4r-5r (f. 4v). Torporley says that the *linea nutus* of the individual parts of a mass of water are parallel to one another (*Lineas vero nutus partiu[m] singulariu[m] esse parallelas*).

73 PHA, Leconsfield MS 241, VIa, f. 24r.

74 W. Warner, 'De corporum non resultantiam ... effectis ex mutua incidentia oriundis', Northamptonshire Record Office (hereafter NRO), Isham-Lamport MS 3422, V, f. 1r.

75 PHA, Leconsfield MS 241, VIa, ff. 24–25

76 Kalmar, 'Problem of impact', 204 and n. 6.

ancestry', and he points out that the term can be found, for example, in Henri Monantheuil's commentary on the Pseudo-Aristotelian *Quaestiones mechanicae* of 1599.[77] But there is another, earlier, source for the term, and one which seems likely to have been the direct source for Monantheuil's own usage, and that is the *Tractatus de motu*, the only published work of the Genevan lawyer Michel Varro (1542–86).

Varro has long been known to historians of science. As long ago as 1857, the English historian William Whewell recognized Varro's work as 'an anticipation of the doctrine of the Composition of forces', while Alexandre Koyré mentions him in his *Études Galiléennes* as a pre-Galilean example of the principle of the proportionality of the velocity of a moving body in relation to distance, and Stillman Drake noted the close similarity between Varro's line and triangle diagrams and those found in Galileo's working papers.[78] Despite these early signs of interest in Varro's work, apart from an article by Serge Moscovici in 1958,[79] little work was done on his treatise until Michele Camerota and Mario Otto Helbing's edition and translation published in 2000.[80] Camerota and Helbing present Varro's work as 'an important chapter in the history of late sixteenth-century mechanics', and its author as an early precursor of the mathematical natural philosophy of Galileo.

It is not surprising that Pepper was unaware of Varro as a possible source for Harriot and Warner's ideas in 1976 when Varro's work was little known, and Warner's papers on the topic had yet to be discovered. It is these new papers, in fact, which lead me to believe that Varro is the 'missing link' in Harriot's theory. While Warner is not often given to citing the authors he makes use of, and seems to have preferred to make his own exhaustive dialectical considerations of questions rather than citing authorities, in the case of his writings on impacts and collisions we find not one, but two, distinct references to Michel Varro, the first of which links him specifically to the concept of *linea nutus*, when he mentions 'the line of force (w[hi]ch Varro calleth linea nutus)'.[81]

It is not merely a question of borrowing a single term, however. I would argue that Varro's ideas on motion and force had a profound effect on the ways in which Harriot and Warner viewed and presented their findings on mechanics. The idea of the *linea nutus* as a way of geometrically representing

77 Pepper, 'Theory of impacts', 138 and n. 13.
78 W. Whewell, *History of the inductive sciences,* 2 vols. (London, 1857), vol. 2, 8; A. Koyré, *Études galiléennes* (Paris, 1966; reprint, 2001), 89, n. 2; S. Drake, *History of free fall. Aristotle to Galileo* (Toronto, 1989), 3.
79 S. Moscovici, 'Notes sur le "De motu tractatus" de Michel Varro', *Revue d'histoire des sciences et de leurs applications*, 11 (1958), 108-29.
80 M. Varro, *De motu tractatus* (Geneva, 1584). See M. Camerota and M. O. Helbing, *All'alba della scienza galileiana. Michel Varro e il suo De motu tractatus. Un importante capitolo nella storia della meccanica di fine Cinquecento* (Cagliari, 2000). On Varro as the probable source for Monantheuil, see pp. 24–26.
81 W. Warner, NRO, Isham-Lamport MS 3422, III, f. 7v.

the effects of force on bodies is central to Warner and Harriot's theories. As Jon Pepper noted in his 1976 article: 'The implicit principle [of Harriot's theory of impacts] is one of the linearity of independent causes, or rather of their effects. The resolutions along the lines of centres, then, seems to be a fundamental part of the theory'.[82] This is also the case with Warner's definition and theorems, which take the 'linearity' of forces, and their tractability to geometrical analysis as their starting point.

Warner also explicitly acknowledges his reliance on Varro's treatment of the relationship between motion, resistance, and weight. Thus, in what appears to be a note added to his papers entitled *De motu et quiete*, long after their original composition (perhaps after his discussions with Harriot and Hues in 1619), we find the following statement:

> By motive power [*potentia motiua*] is understood the ratio of resistance as (according to Varro, as I remember) resistance is understood as the ratio of magnitude or stability of a moveable body, or rather of the support or both.[83]

We also have a brief but suggestive piece of evidence which indicates that Harriot was reading Varro's work. In a 'Memorandum' written sometime in the 1590s, when he was working on ballistic problems, Harriot listed a number of books including 'Varro', 'My notes of ordinance', and 'Proclus de motu'.[84] While it has been suggested that this was a reference to the Roman author Marcus Terentius Varro, whose extant works deal with agriculture and the history of the Latin language, given the context it seems almost certain that Harriot's book-bag contained Michel Varro's *De motu tractatus*.[85] It would appear from this that Harriot's interest in Varro began with his practical mathematical work on the motion of projectiles and falling bodies (because of Varro's attention to the doctrine of motion in general) and subsequently stimulated his interest in the mechanics of colliding bodies.[86]

82 Pepper, 'Theory of impacts', 140.
83 W. Warner, *De motu et quiete*, NRO, Isham-Lamport MS 3422, Notebook IV, fol. 10r: 'Potentia motiua intenditur vel remittitur, pro ratione resistentiae (quod Varronis vt memine vt) resistentia intenditur vel remittitur pro ratione magnitudinis aut stabilitatis corporis mobilis vel potius firmamenti vel vtriusq[ue].'
84 BL Add. MS 6786, f. 364v. This list is discussed by S. A. Walton, *Thomas Harriot's ballistics*, Durham Thomas Harriot Seminar Occasional Papers, no. 30 (Durham, 1999), 19–21. Although Walton suggests in his paper that Harriot's list refers to a work by the Roman author Marcus Trentino Varro, or is a misspelling of the name of the Swiss Jesuit Sebastian Verro (p. 20), it seems more likely, given the subject matter of the other books on the list, that he is in fact referring to Michel Varro.
85 For the suggestion that 'Varro' was Marcus Terentius Varro, see S. A. Walton, *Harriot's ballistics*, 20.
86 On Harriot's working on falling bodies, see Schemmel, *English Galileo*. Schemmel makes no mention of Michel Varro in this excellent study.

There are other areas of Harriot's legacy that are in urgent need of re-appraisal. One of the most obvious of these is Harriot's optics. Although they have been the subject of a series of ground-breaking essays by Johannes Lohne between the late 1950s and the early 1970s, his optical papers still have much to offer historians of science:[87] his work on refraction (including his often-noted anticipation of Snell's law),[88] chromatic dispersion, the location of images, and burning glasses. I do not intend to dwell on Harriot's optics today, however, as I am aware that Robert Goulding, the Harriot lecturer in 2001 is currently working on a book on Harriot and the problem of refraction that will include a detailed examination of Harriot's experimental and mathematical work in this area: Harriot's optics are in safe hands.[89]

Another area that I think would repay further attention is Harriot's en-counter with the great Italian mathematicians of the late sixteenth century, Federico Commandino and Guidobaldo del Monte. Just as Jacqueline Stedall revealed Harriot's debt to Viète in his algebraic work, but also his innovations and independent discoveries, I think a close look at Harriot's engagements with problems inherited from these Italian mathematicians would reveal a similar story. An example that would be well worth pursuing is the question of centres of gravity, a topic investigated both by Harriot and by Warner. It was clearly an area of mathematics that Harriot felt was important, as the inventory of his mathematical papers made by Sir Thomas Aylesbury after this death mentions three bundles of papers on the subject 'De centro gravitatis' (and is the second item of the inventory, immediately after his papers on algebra).[90] The primary impetus here comes from Commandino's *Libro de centro gravitatis solidorum*, published in 1565, which extolled the 'very beautiful' demonstrations of this 'very obscure and very difficult question', which – Commandino informed Cardinal Alessandro Farnese to whom the book is dedicated – offers 'the greatest assistance in clearly understanding many things which are propounded in mathematics'.[91] As we can see from BL Add. MS 6788, Harriot was working

---

87 J. A. Lohne, 'Thomas Harriott (1560–1621), the Tycho Brahe of optics', *Centaurus*, 6 (1959), 113–21; 'Regenbogen und Brechzahl', *Sudhoffs Archiv*, 44 (1965), 401–415; idem, 'Kepler und Harriot: ihre Wege zum Brechungsgesetz', in K. Meyer, F. Krafft, and B. Sticker (eds.), *Internationales Kepler-Symposien weil du Stadt, 1971. Referate und Diskussionen* (Hildesheim, 1973), 187–214.

88 See J. W. Shirley, 'An early experimental determination of Snell's law', *American journal of physics,* 19 (1951), 507–8.

89 For a foretaste of Goulding's forthcoming work see R. Goulding, 'Thomas Harriot's optics, between experiment and imagination: the case of Mr Bulkeley's glass', *Archive for history of exact sciences*, 68 (2014), 137–78.

90 BL Add. MS 6789, f. 448r: 'De Centro Gravitatis, 3 b[undles].'

91 F. Commandino, *De centro gravitatis solidorum* (Bologna 1565), sig. +2r. Harriot was also interested in the work of the Jesuit Luca Valerio (1553–1618) on this topic. See BL Add. MS 6788, f. 358r for a detailed recording of the title page of Valerio's *De centro gravitatis solidorum libri tres* (Rome, 1604). On f. 289r he refers to demonstrations from books 2 and 3 of this work.

his way through the problems in Commandino's book. On folio 262 v, we can see him working on the centre of gravity of a section of a pyramid from p. 35 of Commandino's treatise. However, Harriot is no passive disciple of the Italian mathematician, but a critical one. As we can see from the bottom of this page where he notes: 'This proportion is more suitable, although Commandino did not notice it. Yet it is collected from the elements of his demonstration' (see Figure 2.1).[92] Having grasped the essence of the problem, Harriot is satisfied with Commandino's solution, but presents it in what he thinks is a more elegant form. In another example, also taken from his work on Commandino's book, he criticizes the Italian mathematician's demonstration of a proposition which, he says, is presented 'unclearly, and in another fashion' than the one that Harriot has set down.[93] Harriot's attitude sometimes goes beyond criticism to emulation – in the sense of seeking to ambitiously surpass one's rivals.[94] Thus here we see Harriot having once again formulated a problem (concerning the centre of gravity of a parabolic section) differently from Commandino writing at the bottom of the page 'Vale Comandine tu non habes magisteriu[m]' – which we could render as 'farewell Commandino, you don't have the best solution'![95] Harriot was also working on exactly the same problems – those concerning the centre of gravity of solids – in the *Liber mechanicorum* of Guidobaldo del Monte, published in 1588. But Harriot presents his demonstration in algebraic terms, whereas Guidobaldo restricts himself to a more traditional geometrical style,[96] and when working through Guidobaldo's demonstrations he presents alternative methods of his own.[97]

Another problem that taxed both Harriot and Warner was one taken from Commandino's translation of the work of Pappus of Alexandria, the *Mathematicae collectiones*, also published in 1588.[98] This is the section, or 'resection' of space (*De resectione spatii*), which had been handled by Pappus

---

92 BL Add. MS 6788, f. 262v.: 'Haec ratio est magis accommoda quam Comandinus obseruavit. Attamen ex elementis apud illu[m] colligitur.' Cf. Commandino, *De centro gravitatis solidorum*, 35.

93 PHA, HMC MS 241 VIa, ff. 14r and 15r.: 'comandinus de centr[o] grav[itatis] prop. 22, sed obscurè et alio modo'.

94 See, for example, his executive summary of ways to ascertain the centre of gravity in sections of triangles, parabolas, cones, and pyramids, which are 'much easier to use', Add. MS 6788, f. 321r: 'De centro grauitatis frustoru[m] eclogae aliter dispositae, vsui magis accomoda'.

95 BL Add. MS 6788, f. 332v. Harriot repeats this formula on f. 335v, where he gives his solution to the centre of gravity of a conic section.

96 See e.g., BL Add. MS 6788, f. 267v: 'De centro gravitatis Trapezij vt Archimedis. Guidus Vbaldus, pag. 108'. Cf. G. del Monte, *In duos Archimedis aequeponderantium libros paraphrasis* (Pesaro, 1588), 108.

97 See, e.g., BL Add. MS 6788, f. 283r ('Aliter. et nostro methodo').

98 F. Commandino, *Pappi Alexandrini mathematicae collectiones à Federico Commandino Urbinate in latinum conversa, et commentariis illustrata* (Pesaro, 1588).

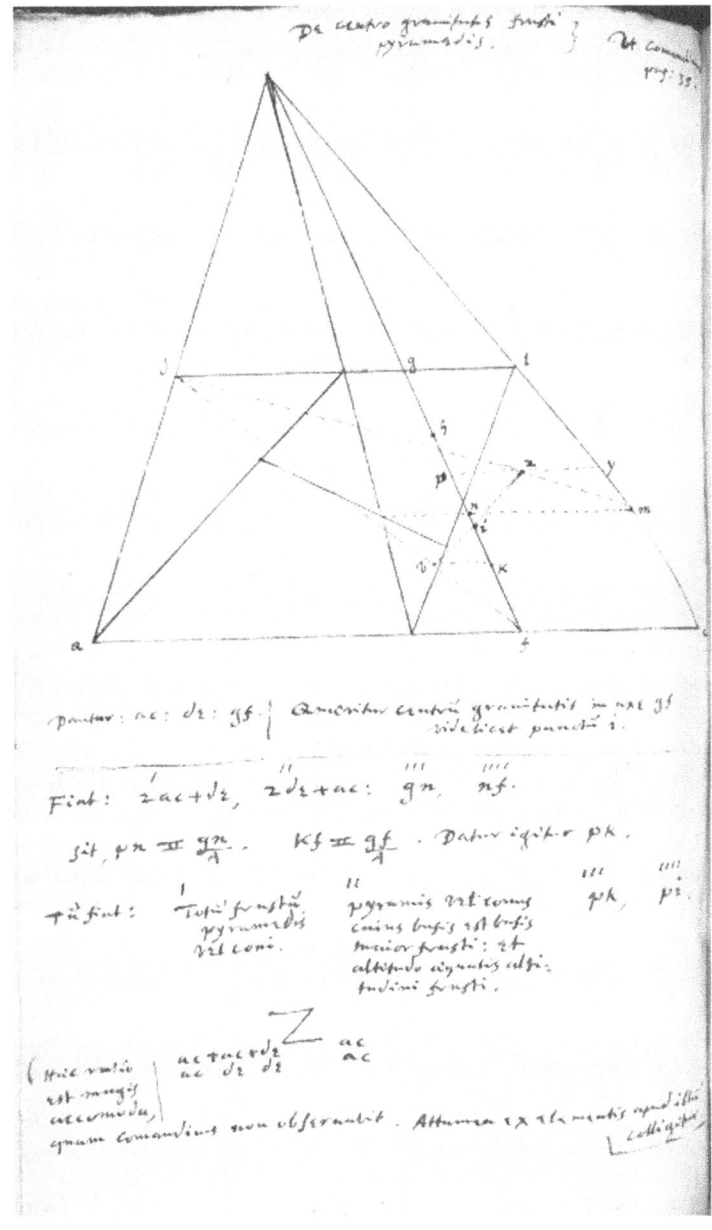

Figure 2.1 Harriot, page of notes on Federico Commandino's *De centro gravitatis solidorum* (1565). London, British Library, Add. MS 6788, f. 262v. By permission of the British Library.

in two books. This involved the solution of a problem in which the geometer is given two straight lines and a point in each, and is required to draw a third line through a third given point so that the rectangle contained by the two intercepts are equal to a given rectangle.[99] As Jacqueline Stedall noted in 2002, this is 'a topic treated more than once by Harriot and found several times amongst Warner's papers', and an inventory of Warner's papers made by Herbert Thorndike after Warner's death includes an entry for a bundle of papers on this topic.[100] In fact, we know that Warner and Harriot shared ideas on the topic as Warner made fair copies of Harriot's solutions to some of these problems, marked 'T.H.' at the top left-hand corner.[101] Harriot and Warner also worked on other problems from Pappus, including determinate sections, and tangents.[102] A short manuscript treatise by Warner presenting ten problems in tangency can be found in the hand of Robert Payne in Chatsworth House.[103]

It is important that we try to piece together these engagements with Italian Renaissance mathematics, I think. Not simply in order to chart its reception in the English context – although that would not be without interest – but because we know that Harriot used these new mathematical tools when confronting physical questions which interested him. As I said earlier, Schemmel has shown how Harriot was able to use specific problems from Archimedes and Pappus when he was wrestling with the nature of the ballistic trajectory. A case in point can be found in a cryptic passage amongst his papers on centres of gravity in BL Add. MS 6788[104] (See Figure 2.2). We can see from this passage that centres of gravity are not mere abstractions for Harriot but concern real physical bodies and their behaviours. The diagram and the accompanying text suggest that the centre of gravity of any particular moving object is not in the object itself (which is 'the subject of gravity'), but – following Aristotle – at the centre of the earth ('centrum terrae').[105] How we interpret the other remarks here 'In the descent of a sphere along a plane – whether it is revolved or not – motion is caused of necessity, but not the end [of motion]. Double contact', is not

99 Commandino, *Mathematicae collectiones*, p. 158b: 'Per datam punctum rectam lineam ducere secantem à duabus rectis lineis positione datis ad data puncta, lineas, quae spacium contineant dato spacio aequale.'

100 Stedall, *English algebra*, 242, n. 89. An example of Warner working on the resection problem is BL Add. MS 4396, ff. 40r–44r.

101 NRO, Isham-Lamport MS 3422, VI, fol. 12: 'De resectione rationis'; ff. 18–22r: 'De resectione spacij'.

102 Ibid., ff. 14–15: 'De determinata sectione' and ff. 4–6: 'De determinata sectione'.

103 W. Warner, 'De tactionibus', Chatsworth House Archive, Hobbes MS B.5. See also another non-autograph copy of this short work in BL Harley MS 6755, ff. 3–14.

104 BL Add. MS 6788, f. 344v.

105 'Centru[m] grauitatis no[n] in subiecto grauitatis. Motus necessariò causatur sed non determinatio in des[c]ensu sphaerae p[er] planu[m] an reuoluatur necne. Duplex contactus.'

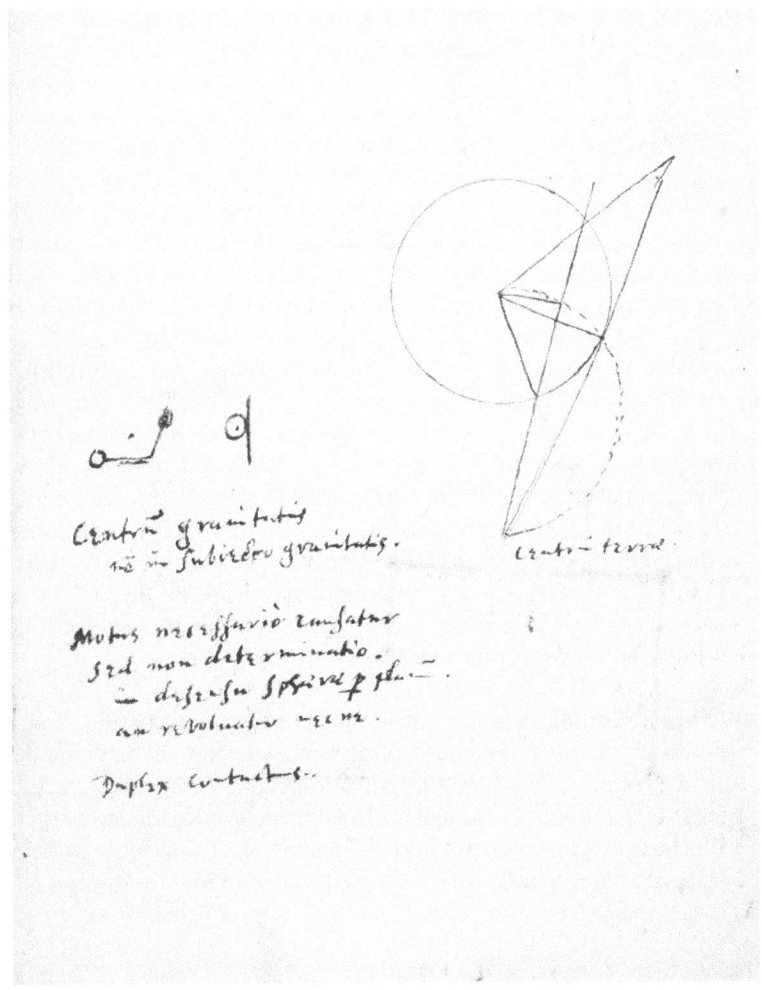

Figure 2.2 Harriot's notes on the centre of gravity and the motion of a sphere along an inclined plane. London, British Library, Add. MS 6788, f. 344v. By permission of the British Library.

immediately obvious, but it suggests that more careful attention to Harriot's doctrines of motion and mechanics might be enhanced by a fuller understanding of these mathematical problems.[106]

---

106 I suspect that Harriot is responding here to Hero of Alexandria's discussion of a sphere rolling down an inclined plane in the *Mechanica*. See S. Roux and E. Festa, 'The enigma of the inclined plane from Hero to Galileo', in W. R. Laird and S. Roux (eds.), *Mechanics and natural philosophy before the Scientific Revolution* (Dordrecht, 2008), 195–221.

A theme arising out of a study of the works of these Italian mathematicians, as well as the works of other continental mathematicians such as Simon Stevin and Marino Ghetaldi, which I raised in my first Harriot lecture, but which has still not been sufficiently exhausted, is the topic of hydrostatics, or statics more generally. Sir William Lower felt that Harriot had allowed Ghetaldi to beat him into print on the question of the Archimedean method of calculating specific gravities by weighing objects in water. This was certainly an area that interested both Harriot, and other members of the Northumberland household. Warner's papers included a work on the statics of solid and fluid bodies, the *Ad praxim staticam elementa quaedam accomoda*,[107] and Nathaniel Torporley wrote a brief critique of Simon Stevin's *Hydrostatica*, 'De pondere aquae', at the request of Henry Percy, in which he defends several Archimedean propositions, and seeks to show that Stevin had fallen into a number of serious errors.[108] It is interesting to note, by the way, that Torporley, like Harriot and Warner, makes use of the *linea nutus* concept in this work to explain water pressure.[109] An assessment of Harriot's specific gravity experiments, in the wider context of the engagement with statics and hydrostatics in the Northumberland would repay the effort, I suspect.[110]

Another area which could usefully be revisited is Harriot's work on spirals and rhumb lines. Harriot's colleague Robert Hues, in the preface of his *Tractatus de globis et eorum usu* in 1594 expected the imminent publication of Harriot's thoughts on this subject:

> We are awaiting a whole treatise on the generation, nature, and use of rhumbs by Thomas Harriot, who is most expert in mathematics and universal philosophy. In which many things concerning this argument have been subtly and acutely thought out, elaborated with great industry, refined with the utmost judgement, and weighed in the balance of mathematical demonstrations: which work we hope to see published very soon.[111]

---

107 BL Harley MS 6754, ff. 2–74, and NRO, Isham-Lamport MS 3422, I.

108 N. Torporley, 'De pondere aquae quo premuntur ij quibus altius incumbit. Quaestio D[omi]no Henrico Comite Northumbriae proposita, et ventiliata', BL Add. MS 4458, ff. 4r–5r.

109 See note 72 above.

110 A recent contribution to our understanding of Harriot's hydrostatics is provided in N. Biggs, 'Thomas Harriot on the coinage of England', *Archive for history of exact sciences*, 73 (2019), 361–83. See also S. Clucas, '"The curious ways to observe weight in water": Thomas Harriot and his experiments on specific gravity', *Early science and medicine*, 25 (2020), 302–27.

111 R. Hues, *Tractatus de globis et eorum vsu* (London, 1594), 111.: 'De rumborum ortu, natura & vsu integrum tractatum expectamus à Thomas Harioto matheseos & vniuersae Philosophiae peritissimo. à quo in hoc argumento multa subtiliter & acutè excogitata, magnâ industriâ elaborata, summo iudicio expolita sunt, & ad Geometricarum demonstrationum trutinam perpensa: quem propediem edendum speramus'. See D. B. Quinn and J. W. Shirley, 'A contemporary list of Harriot references', *Renaissance quarterly*, 22 (1969), 9–25 (13–14).

Hues had every right to be so expectant – Harriot had done a lot of work. In an inventory of his manuscripts made after his death, we find listed 'A black box full of papers of Rhombes'.[112] While we have the excellent and comprehensive article on Harriot's work on rhumb lines and meridional parts by Jon V. Pepper,[113] might it not be possible to reconstruct a treatise on this topic, beginning with those papers headed 'De rumbis' and 'De helicis', in HMC 240, II? Pepper published a selection of the most significant manuscript pages in his 1968 article, but he noted at the time that his transcriptions were 'only a small selection of those available'.[114] Anyone who wished to undertake this should begin with Pepper's article as he has provided any future editor with a helpful calendar of the relevant folios they would need to piece together.[115]

Schemmel's historical epistemological method strikes me as a very fruitful approach to follow in other areas than Harriot's mechanics. Comparison with other mathematicians and natural philosophers in the late sixteenth and early seventeenth century and how they dealt with the same kinds of problems will help us to map out the problematic areas in which mathematics and natural philosophy overlap. Historians of astronomy, as Jim Bennett noted at the beginning of his 1995 lecture, 'seem often to be disappointed by Thomas Harriot'.[116] But what if we were to set him in a wider European context of astronomical observation? If we look at other comparable figures in Europe at Harriot's time, such as Nicolas-Claude Fabri de Peiresc (1580–1637), what can we learn about observational practice? If we look, for example, at the famous astronomical observations made by Peiresc together with his friend Joseph Gaultier de la Vallette (1564–1647) on 24 November 1610. The two men – like Harriot – were observing the moons of Jupiter (which they refer to here as 'the Medicean planets' – *les Planètes Medicées*) when they became aware of 'a certain little illuminated cloud composed of two stars' in the middle of Orion,[117] making them the

---

112 BL Add. MS 6789, f. 449r.
113 J. V. Pepper, 'Harriot's calculation of the meridional parts as logarithmic tangents,' *Archive for history of exact sciences*, 4 (1968), 359–413.
114 Pepper, 'Meridional parts', 391.
115 Pepper, 'Meridional parts', 391–93.
116 Fox (ed.), *Elizabethan man of science*, 137.
117 N. C. Fabri de Peiresc, *Journal des observations*, Bibliothèque Inguimbertine, Carpentras, France, MS 1803, f. 189r: 'In Orione media... ex duabus stellis composita nubecula[m] quamdam illuminat[am].' For a recent transcription of the relevant manuscript see H. Siebert 'De Peirescs Nebel im Sternbild Orion - eine neue Textgrundlage für die Geschichte von M42', *Annals of science*, 66 (2009), 231–46 (239). See also H. Siebert, 'Die Entdeckung des Orionnebels. Historische Aufzeichnungen aus dem Jahr 1610 neu gesichtet', *Sterne und Weltraum*, 11 (2010), 32–42, and S. L. Chapin, 'The astronomical activities of Nicolas Claude Fabri de Peiresc', *Isis,* 48 (1957), 13–29 (19). Harriot's 'first obseruation... of the newfound planets about Iupiter' was dated 17 October 1610. PHA, HMC MS 241 IV, f. 3r.

first observers of the Orion nebula (now known as M42). While this was a momentous discovery, what I find interesting here is the similarity of Peiresc's observational practices to those of Harriot. Like Harriot, Peiresc frequently observed with others (in this case Gaultier), and like Harriot he recorded their shared doubts about what they had seen through the telescope. Peiresc names the four Galilean moons 'Francisc[us]', 'Ferdin[andus]', 'Cosmi ma[ioris]' and 'Cosmi min[oris]' after the four recent Medicean Grand Dukes of Tuscany.[118] Peiresc records the varying levels of certainty about the appearance of these remote planetary bodies. At the top of the page under the date he writes 'Monsieur Gaultier began to see the Medicean planets' ('M. Gaultier a com[m]ence à voir les Planetes Medicées'). Further down he writes 'we began to see them ourselves in this way' ('Nous auons com[m]encé à les voir nous-memes En ceste sorte'.). He then notes the differing status of these observed bodies, there is no doubt about the first ('de prima non ambigit [ur]'), the second was doubted very much ('de 2^a dubitatur valde') although the third less so ('de 3^a minor est dubita[ti]o'). This use of multiple witnessing is typical of Harriot too, who often recorded his own doubts and those of his companions Christopher Tooke or Nicholas Sanders about particular observations.[119] A more sustained comparison of the observational methods of Peiresc and Harriot could well prove fruitful.[120] Cometography is another area where comparisons with broader trends would be instructive. Harriot made careful observations and measurements of both the 1607 and 1618 comets.[121] Tracking down the observations of his contemporaries both in print and manuscript would allow Harriot's work to feature as part of a broader investigation of the early seventeenth-century understanding of these astronomical phenomena.

These are just a few suggestions for where future research on Harriot might be directed. Harriot scholarship is moving into a new and exciting phase. Much has already been accomplished, but much remains to be done.

---

118 Cosimo I de' Medici (d. 1574), Francesco de' Medici (d. 1587), Ferdinando de' Medici (d. 1609) and Cosimo II de'Medici (d. 1621). I would like to thank Prof. Dr. Harald Siebert for his advice regarding Peiresc's observations.

119 See, e.g., PHA, HMC 241 IX, f. 3r (9 April 1611, lunar observations with Tooke and Sanders); PHA, HMC 241 VIII, f. 10r. (26 January 1611/12, sunspot observations with Tooke).

120 Although Harriot's and Galileo's maps and drawings of the moon have been much discussed, as far as I know, Peiresc's moon drawings have not been discussed in this context. On Harriot and Galileo, see T. F. Bloom, 'Borrowed perceptions: Harriot's maps of the Moon', *Journal for the history of astronomy*, 9 (1978), 117–22; H. Bredekamp, 'Gazing hands and blind spots: Galileo as draftsman', in J. Renn (ed.), *Galileo in context* (Cambridge, 2001), 153–92, esp. 176–84; S. Pumfrey, 'Harriot's maps of the Moon: new interpretations', *Notes and records of the Royal Society*, 63 (2009), 163–68; A. Chapman, 'A new perceived reality: Thomas Harriot's Moon maps', *Astronomy & geophysics*, 50 (2009), 1.27–1.33.

121 See PHA, HMC MS 241 VII, ff. 1–39.

Thanks to the efforts of Jacqueline Stedall, Matthias Schemmel, and Robert Goulding, the dream of an edition of Harriot's papers once cherished by John Shirley, David Quinn, and Cecily Tanner is becoming a reality. Although the transcriptions and commentaries are not yet complete, the process has started. Eventually a new generation of scholars will come to Harriot's papers online, with expert commentary to guide them. I am sure that when these new scholars sift through these fascinating papers, they will find quite another Harriot than the one that I have found, a different Harriot from the one presented by Quinn and all the Harriot lecturers who came after him: they will find a Harriot waiting to be asked new, and as yet unasked questions.

Acknowledgement

All images from BL Add. MS 6788 in this article are taken from the ECHO Thomas Harriot Online website currently hosted by the Max Planck Institute, <http://echo.mpiwg-berlin.mpg.de/content/scientific_revolution/harriot>. These images have been classified for use under the Creative Commons license.

3 'Our learned countryman'. Thomas Harriot and the Emergence of Mathematical Community in Seventeenth-Century England

Philip Beeley

Harriot's standing among English mathematicians

For the great seventeenth-century Oxford mathematician John Wallis (1616–1703) there was good reason for naming Galileo Galilei (1564–1642) and Thomas Harriot (1560–1621) in the same breath. Writing in November 1670 to the Grand Duke of Tuscany (1642–1723), Wallis recalled Cosimo III de' Medici's ceremonial visit to Oxford – when still as crown prince – more than a year earlier.[1] Besides going to the Bodleian Library, visiting a number of colleges, including Queen's, Magdalen, and Christ Church, Cosimo had been entertained by an architectural lecture. This had been given by Wallis and concerned his mathematical model for a free-standing ceiling that Christopher Wren (1632–1723) had taken up for the Sheldonian Theatre, which at that time, in the summer of 1669, was still being built.[2]

In his letter, Wallis praises Florentine architecture as being preeminent in Europe, and he praises Galileo, 'the great mathematician of a great duke, whom I venerate as the father of the new philosophy'.[3] He proceeds to talk of other Italian mathematicians such as Gerolamo Cardano (1501–76), Niccolò Tartaglia (1499–1557), and Bonaventura Cavalieri (1598–1647), before turning to François Viète (1540–1603), the French mathematician who further developed the ancient Greek algebraic tradition of Diophantus and introduced arithmetica speciosa or literal arithmetic into modern discussion. Now on the topic of algebra, Wallis could cut to the chase, for Viète, he tells Cosimo, had found worthy successors in two English scholars,

---

1 On Cosimo's visit, see Anna Maria Crinò (ed.), *Un principe di Toscana in Inghilterra e in Irlanda nel 1669: relazione ufficiale del viaggio di Cosimo de' Medici tratta dal 'Giornale' di L. Magalotti, con gli acquerelli palatini* (Rome, 1968), 101–4.

2 See Thomas Lamplugh to Joseph Williamson, 5/[15] May 1669, in Philip Beeley and Christoph J. Scriba (eds.), *The correspondence of John Wallis (1616–1703)*, 4 vols. (to date), (Oxford, 2003–14), vol. 2, 169–71.

3 John Wallis to Cosimo III de' Medici, 9/[19] November 1670, in Beeley and Scriba (eds.), *Correspondence of John Wallis*, vol. 3, 401–5 (402): 'Certe Galilaeum Vestrum (Magnum Magni Ducis Mathematicum) tacere non debeo; quem ut Novae Philosophiae Parentem veneror'.

DOI: 10.4324/9781003096580-4

William Oughtred (1575–1660) and Thomas Harriot. He was careful to prefix both of these names with 'our' in order to counterbalance the many occurrences of 'your' in the letter which literary etiquette had demanded.[4] England, so the general tone of the letter, was now up there with the best. But equally, if not more important for Wallis were remarks which he makes specifically in relation to Harriot: that from him René Descartes (1596–1650) had derived the principles of his geometry, but without bothering to mention the Englishman's name.

By the time of Wallis's letter to Cosimo, Oughtred and Harriot had come to represent the openness and modernity of mathematics in England, the two men were seen in many ways as the figureheads of the country's emerging mathematical community. In countless publications covering the whole spectrum of mathematics, from tracts on accountancy, dialling or general arithmetic up to learned treatises on the centre of gravity or algebra we find references to 'our famous countryman Mr Oughtred' or 'that learned mathematician of our Nation Mr Harriot', and the like.[5]

The importance of the role ascribed to Oughtred and Harriot is to be seen in the context of a remarkable growth of the mathematical sciences in England particularly during the second half of the seventeenth century. This progress was reflected not only in the newly established Royal Society of

---

4 Ibid., 402–3: 'Quem feliciter secutus sunt Oughtredus noster; et Harriotus item noster, ex quo Cartesius (celato nominee) praecipua suae Geometriae Fundamenta mutuatus est'.

5 See for example Richard Balam, *Algebra: or, the doctrine of composing, inferring, and re-solving an equation* (London: J. G. for R. Boydell, 1653), 130: 'All kindes of mixt Equations may be resolved by one generall way, which is taught by two learned men of the English Nation. Namely Mr. Will: Oughtred in his Clavis Mathematicae, and Mr. Thomas Harriot in his Artis Analyticae Praxis'; John Kersey, *The elements of that mathematical art commonly called algebra* (London: William Godbid for Thomas Passinger, 1673), preface, Sig. b2v–b3r: 'But the Excellency of the Algebraical Art is best known to those that are acquainted with the most eminent Writers upon that Subject; among which, these are deservedly Famous [...] Cardanus, Tartaglia, Clavius, Stevinus, Vieta [...], Mr. William Oughtred, (our learned Countrey-man,) whose Clavis Mathematicae, for Solid matter, neat Contractions, and succinct Demonstrations, is hardly to be parallel'd, Mr. Thomas Harriot, (another learned Mathematician of our Nation,) Ghetaldus, Andersonus, Bachetus [...]'; Edward Sherburne, *The Sphere of Marcus Manilius made an English poem: with annotations and an astronomical appendix* (London: Nathanael Brooke, 1675). Brooke's Letter to the reader, pp. 2–3: 'On the like Reasons we may conceive we want the many learned Algebraical Works of our famous Countryman Mr. Thomas Harriot, (and of Mr. Warner, into whose Hands they fell) who is esteemed by some of the most knowing Persons alive to have been much Superiour to all that ever writ'; John Ward, 'Synthesis et Analysis vulgo Algebra' [London, 1695]. Advertisement: 'The most knotty and difficult Problems, which are to be met withal in the Works of Euclid, Archimedes, Diophantus, Pergaeus, Vieta, &c. and in those of our own Country-men, as the Learned and Reverend Mr. Oughtred, Mr. Harriot, Dr. Pell, Mr. Kersey, Dr. Wallis, and others; are hereby Explained, Demonstrated, and made Intelligible to mean Capacities'; Joseph Raphson, *Analysis Æquationum universalis* (London: Abel Swalle, 1690), p. 2: 'Jam vero ad (g) seu primum membrum inveniendum, signetur æquatio punctis, modo, quem docuere Vieta, nostratesque Harriotus & Oughtredus, tum potestas resolvenda, Tum Coefficientes';

London, but also in an increasingly sophisticated level of practical mathematics in accountancy, commerce, navigation, and instrument making. New mathematical learning permeated workshops, warehouses, dockyards, was a topic of conversation in coffee houses and taverns, and was disseminated by means of printed books, manuscripts, journals, and letters.[6] While English printers and booksellers were notoriously reluctant to produce mathematical books of a more theoretical nature – not least because of expense of typesetting and the relatively small print-runs – there was a ready and burgeoning market for more practical books. As John Kersey (c.1616–77) noted in 1673

> Nor are Arithmetick and Geometry excellent in themselves only, but highly esteem'd also for their manifold Utility, as well in the Employments of Men about Accompts, Trade, Building, Measuring of Land, and divers other common Affairs, as in facilitating and enlivening divers other Noble Arts, for how can Harmonical Composition in Musick, or exact Measure and Proportion in Painting be perform'd, without the assistance of Arithmetick and Geometry.[7]

Kersey, a largely self-taught teacher of mathematics in London, published his *Elements* with the support of his friend, the indefatigable promoter of all things mathematical, John Collins (1626–83), who took more than five years to persuade booksellers to undertake the project.[8] Although Kersey draws heaviest on the English authors Harriot, Oughtred, Wallis, and Isaac Barrow (1630–77), this is probably more for reasons of the accessibility of their works, some of which were written in English rather than the learned Latin language. But he cites French authors, too, such as Descartes and Florimond Debeaune (1601–52), the Dutchmen Frans van Schooten (1615–60) and Jan Hudde (1628–1704), Italian mathematicians such as Cardano and Tartaglia, and the Ragusan author Marino Ghetaldi (1568–1626), in a well-balanced and matter-of-fact way without any kind of national point scoring. The declared aim of Kersey's work was 'to give such of my Mathematical Countreymen as are altogether strangers to, and desirous to be acquainted with the so much celebrated Art called Algebra, a plain and intelligible Introduction to its Doctrine, as also a considerable taste of its Use'.[9] And this laudable aim of Kersey's corresponds to his remarkably well-tempered approach.

---

6 See Philip Beeley, 'Practical mathematicians and mathematical practice in later seventeenth-century London', *British journal for the history of science*, 52 (2019), 225–48.

7 Kersey, *Elements of that mathematical art*, preface, Sig. b2r.

8 See Philip Beeley, 'Advancing the "Analytick doctrine": the making of John Kersey's Elements of Algebra', in Philip Beeley and Ciarán Mac an Bhaird (eds.), *Mathematical book histories. Printing, provenance, and practices of reading* (Basel, forthcoming 2022).

9 Kersey, *Elements of that mathematical art*, Sig. b3r.

When 12 years later John Wallis brought out the next substantial English publication on the topic, his *Treatise of algebra, both historical and practical* (1685),[10] it marked the end of an equally long period of gestation as that of Kersey's *Algebra*, having originally been conceived some 20 years earlier. But Wallis's readers met with quite a different kind of presentation to that of the London practitioner. Ostensibly an account of the origins, progress, and advancement of algebra, the Oxford mathematician's treatment was anything but balanced. Although he went to some considerable lengths to set out the ancient Greek and Arabic sources and to discuss the medieval and Renaissance contributions to the development of algebra, by far the largest sections are devoted to presenting in extenso Harriot's theory of equations, drawn from the *Artis analyticae praxis*, as this was published by Walter Warner (1563–1643) in 1631,[11] with shorter sections on Oughtred's *Clavis mathematicae* (first published 1631 and reprinted often since),[12] and John Pell's *Introduction to algebra* (1668).[13] But while the *Treatise of algebra* undoubtedly had the merit of setting out key sections of Harriot's *Artis analyticae praxis* in a systematic manner – something Kersey's work did not even aspire to do – Wallis leaves out no opportunity to seek to further substantiate his conviction set out in his letter to Cosimo, that Descartes had drawn much of his seminal work, the 1637 publication *La géométrie* from Harriot, but without naming his source.[14]

10 John Wallis, *A treatise of algebra, both historical and practical. Shewing, the original, progress, and advancement thereof, from time to time; and by what steps it hath attained to the heighth at which now it is* (London: John Playford for Richard Davis, 1685).

11 *Artis analyticae praxis, ad æquationes algebraicas nova, expedita, & generali method, resolvendas: tractatus e posthumis Thomae Harrioti philosophi ac mathematici celeberrimi schediasmatis summa fide & diligentia descriptus* (London: Robert Barker, 1631).

12 [William Oughtred], *Arithmeticae in numeris et speciebus institutio: quae tum logisticae, tum analyticae, atque adeo totius mathematicae, quasi clavis est* (London: Thomas Harpur, 1631).

13 [John Pell], *An introduction to algebra, translated oat of High-Dutch into English, by Thomas Brancker* (London: W. G. for Moses Pitt, 1668).

14 Wallis, *Treatise of algebra*, preface, Sig. a4r: 'In sum, He [sc. Harriot] hath taught (in a manner) all that which hath since passed for the Cartesian method of Algebra; there being scarce any thing of (pure) Algebra in Des Cartes, which was not before in Harriot; from whom Des Cartes seems to have taken what he hath (that is purely Algebra) but without naming him'. The Savilian Library exemplar of *Artis analyticae praxis*, Oxford, Bodleian Library, shelf-mark Savile O 9, formerly in the possession of Robert Payne, has a sheet inserted containing handwritten notes by Wallis on Harriot. In these notes, dated 27 March 1677, he makes a similar claim: 'This treatise of Mr Harriots, was, (it seems) so well liked by Des Chartes, that he hath, in a manner, transcribed the whole of it for the substance (though in other order & words) into his Geometry (but without so much as ever naming the Author,) which was first published in the year 1637 (in French) six years after this was first extant'. In reply to a question from Samuel Morland, Wallis would later deny that he ever accused Descartes of plagiarism, asserting instead only that much if not all of Descartes's algebra had been treated of before by others, 'especially by our Harriot', but without their being named. See John Wallis to Samuel Morland, 12/[22] March 1688/9, in John Wallis, *Opera mathematica*, 3 vols. and suppl. (Oxford: at the Sheldonian, 1693–99), vol. 2, 207–13. See also Christoph J. Scriba, 'Wallis und Harriot', *Centaurus*, 10 (1964), 248–57; John W. Shirley, *Thomas Harriot. A biography* (Oxford, 1983), 10–12.

In order to understand why Wallis might have been convinced of Descartes's wrongdoing, it is important to note that he only came across Harriot's *Artis analyticae praxis* sometime after he had worked systematically through Descartes's mathematical work. It was the superficial similarities between Harriot and Descartes which led Wallis to his view, not a serious attempt to derive the one from the other. But two contemporary episodes also contributed to Wallis's false assertion of plagiarism, and one of these he recites with obvious delectation in his *Treatise of algebra*. During a discussion in Paris around 1648, the English natural philosopher and mathematician Charles Cavendish (1594–1654), then living in exile, showed Gilles Personne de Roberval (1602–75), professor of mathematics at the Collège de France, a copy of Harriot's *Artis analyticae praxis*. According to Cavendish, Roberval, too, had quickly come to recognize Descartes's supposed dependence on Harriot:

> I admire (saith M. Roberval) that Notion in Des Cartes of putting over the whole Equation to one side, making it equal to Nothing, and how he lighted upon it. The reason why you admire it (saith Sir Charles) is because you are a Frenchman; for if you were an Englishman, you would not admire it. Why so? (saith M. Roberval.) Because (saith Sir Charles) we in England know whence he had it; namely from Harriot's Algebra. What book is that? (saith M. Roberval,) I never saw it. Next time you come to my Chamber (saith Sir Charles) I will shew it you. Which a while after, he did: And upon perusal of it, M. Roberval exclaimed with Admiration (Il l'a veu! Il l'a veu!) He had seen it! He had seen it! Finding all that in Harriot which he had before admired in Des Cartes; and not doubting but that Des Cartes had it from thence.[15]

Another clearly related incident took place during Descartes's last visit to Paris in the summer of 1648. An account of it was conveyed to the English public by the Roman Catholic playwright William Joyner (1622–1706) on his return to England from France later that year. According to Joyner, Roberval at a public meeting accused Descartes to his face of having taken most of his algebraic ideas from Harriot. This attack is corroborated from other sources too, and is a reflection of the deep dislike and indeed open hatred with which the author of the *Géométrie* was seen by many in the mathematical community in Paris. But Wallis and his friends in England remained blissfully unaware of how

---

15 Wallis, *Treatise of algebra*, 198. This episode is also recounted in Adrien Baillet, *La vie de Monsieur Des-Cartes*, 2 parts (Paris: Horthemels, 1691), part 2, 540–1. See also Jacqueline A. Stedall, *The greate invention of algebra. Thomas Harriot's treatise on equations* (Oxford, 2003), 28–30.

the existence of such infighting in scientific circles across the Channel might have coloured their evidence.[16]

The promise of Harriot's papers

It is of course a pertinent question to ask how much Wallis and his contemporaries knew of the work of Harriot and Oughtred. Wallis's first view of the *Artis analyticae praxis* was cursory, as already mentioned, but before he came to write corresponding sections of the *Treatise of algebra* he had had the time to study Harriot's one published mathematical tract in detail.[17] We cannot date this detailed study precisely, but circumstantial evidence suggests that it was carried out in the late 1660s, when much of his *Treatise of algebra* was completed.[18] There are also grounds for believing that Wallis had access to some of Harriot's unpublished manuscripts at this time, for in the *Treatise of algebra* he cites Harriot's treatment of negative roots although nothing on this topic is to be found in the published tract.[19] Furthermore, Wallis actually mentions having earlier seen some of Harriot's unpublished manuscripts in notes which he wrote some years later, in 1677, and which were inserted in the copy of *Artis analyticae praxis* deposited contemporaneously in the library of the Savilian professors in Oxford (see Figure 3.1):

16 On the background to Wallis's accusation of Descartes's plagiarism of Harriot, see Philip Beeley and Christoph J. Scriba, 'Wallis, Leibniz und der Fall von Harriot und Descartes. Zur Geschichte eines vermeintlichen Plagiats im 17. Jahrhundert', *Acta historica Leopoldina*, 45 (2005), 115–29.

17 Wallis points to this more detailed study, continuing largely from where Payne had left off, in notes inserted in the Savilian exemplar of *Artis analyticae praxis*, Oxford, Bodleian Library, shelf-mark Savile O 9: 'For, coming toward the middle of the book, & so onwards, I find many mistakes uncorrected: some of which I have amended with the pen, & some I have noted in loose peeces of paper, put between the leaves in their proper places, to avoid defacing the book. And, particularly, I have added what was wanting in the limitation of the 19th, 20th, & 21th propositions of the Third section, pag. 45, 46. the want of which, made the publisher to give us that Note of his Hesitation, pag. 46'.

18 Acting as Wallis's agent in London, John Collins sought without success from the late 1660s to find a publisher willing to take on his algebra treatise. See, for example, Collins's letter to Francis Vernon, mid-December 1671, Cambridge University Library, MS Add. 9597/13/5, f. 68r–69v: 'Dr Wallis hath wrote many things of algebra, which he would have printed, with his former works, in Holland, (seeing he cannot get it done here,) but I think they will not comply'.

19 Wallis, *Treatise of algebra*, 128: 'Beside the Positive or Affirmative Roots, (which he doth, through his whole Treatise, more especially pursue, as the principal and most considerable:) He takes in also the Negative or Privative Roots; which by some are neglected'. Johannes A Lohne draws attention to this discrepancy in his 'Dokumente zur Ravalidierung von Thomas Harriot', *Archive for history of exact sciences*, 3 (1966), 185–205 (193-94). See also Muriel Seltman, 'Harriot's algebra: reputation and reality', in Robert Fox (ed.), *Thomas Harriot. An Elizabethan man of science* (Aldershot and Burlington, VT, 2000), 153-85; Stedall, *Greate invention of algebra*, 30–31.

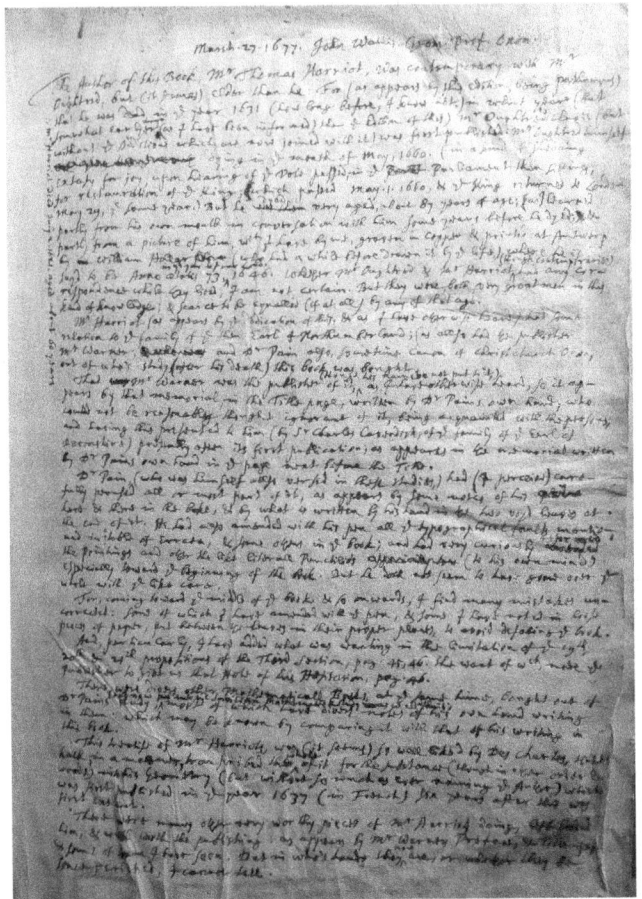

Figure 3.1 Autograph notes by John Wallis on Harriot, prefixed to the Savilian Library copy of the *Artis analyticae praxis*. Bodleian Library, shelf-mark Savile O 9. By permission of Bodleian Libraries, University of Oxford.

There were many other very worthy pieces of Mr Harriots doing, left behind him, & well worth the publishing: as appears by Mr Warners Preface, & Title-page & some of them I have seen. But in who's hands they now are, or whether they be since perished, I cannot tell.[20]

The most likely route to Harriot's papers at this time would have been through John Pell (1611–85). Alongside Wallis, Pell was probably the only contemporary English mathematician capable of assessing the true value of

---

20 *Artis analyticae praxis*, Oxford, Bodleian Library, shelf-mark Savile O 9.

Harriot's mathematical writings, but through his circle of friends he also had the opportunity to study them in depth. There is also written testimony that Pell for a time had some of Harriot's papers in his possession. Collins reports having heard from Pell that Harriot's mathematical learning was such 'had he published all he knew in algebra, he would have left little of the chief mysteries of that art unhandled'.[21] Pell would scarcely have made such a claim on the basis of hearsay. Conceivably, he either showed Wallis some of Harriot's manuscript writings or was at least able to inform him of what he had seen. Be that as it may, what is most remarkable in this connection is that Wallis later denied categorically ever having seen any of Harriot's unpublished scientific papers. Spanning this whole period, Wallis was uncertain of the fate of these papers. He mentions this uncertainty already in his 1677 notes and takes up the topic again seven years later in a letter written to John Aubrey (1626-97), but now with quite a different assertion as to his knowledge of them. In the letter, he veritably exudes with happiness at the news he has received from the antiquary that Harriot's papers were with Henry Hyde (1638–1709), the second Earl of Clarendon. The main reason he gives for this reaction is that he had previously never seen anything of Harriot's apart from the *Artis analyticae praxis*: effectively, they were for him unknown treasures.

It is hard to reconcile Wallis's contradictory claims, even allowing for the fact that the Bodleian note was not open to public scrutiny, while his letter to Aubrey potentially was. There was nothing to be gained by Wallis in being untruthful, because he justly acknowledges the author of what he subsequently writes in his *Treatise of algebra*. On the other hand, he did have a genuine interest in preserving England's unpublished scientific heritage, and by emphasizing to Aubrey that he had never caught sight of Harriot's papers he would have served to underscore their scientific value. A decade earlier, Wallis had played a major role in digesting and editing the works of Jeremiah Horrocks (1618-41) after the Royal Society, fearful that the achievements of the Liverpool astronomer might be lost to posterity, had made this one of its earliest major public projects.[22] Over the years, Wallis had heard of Harriot's papers variously being in the hands of Thomas Hobbes (1588–1679), Seth Ward (1617-89), and Pell. In agreement with Aubrey, he believed that they now rightfully belonged in the Library of the Royal Society (see Figure 3.2):

---

21 John Collins to Francis Vernon, mid-December 1671, Cambridge University Library, MS Add. 9597/13/5, f. 68r-69v; Stephen Jordan Rigaud (ed.). *Correspondence of scientific men of the seventeenth century*, 2 vols. (Oxford, 1841), vol. 1, 151-56.

22 See the minutes of the meeting of the Society on 17 February 1663/4, in Thomas Birch, *The history of the Royal Society of London for Improving of Natural Knowledge, from its first rise*, 4 vols. (London: for A. Millar, 1756-57), vol. 1, 386: 'Mention being made of Mr. Horrox's papers concerning celestial observations, Sir Paul Neile promised to produce some of them: and Dr Crowne was desired to write to Mr. Townley, who had a considerable number of Mr. Horrox's papers, to communicate them, in order to their being made public'.

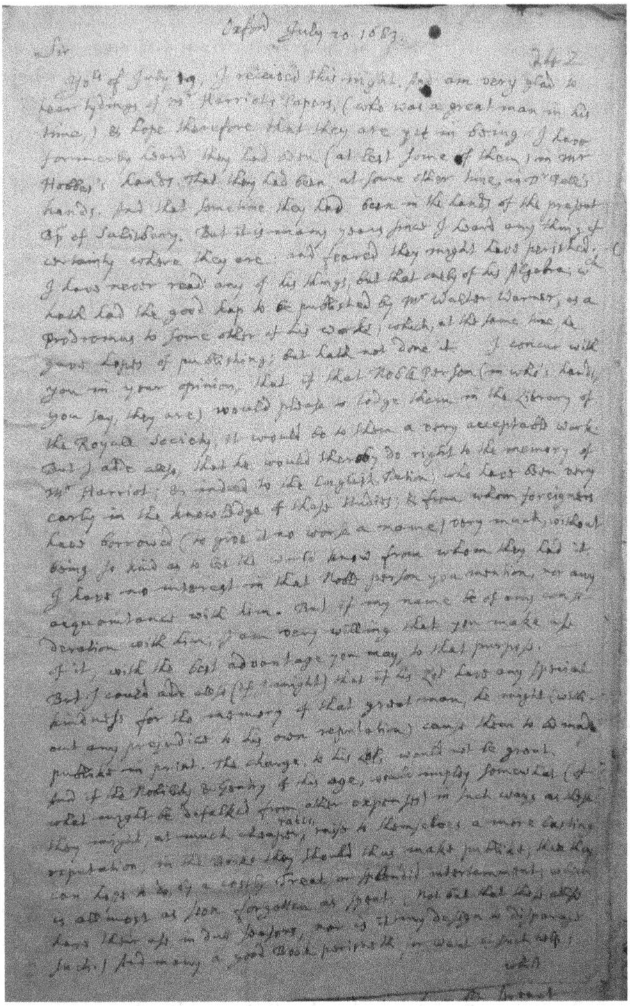

Figure 3.2 Wallis on news of Harriot's papers: John Wallis to John Aubrey, 20/[30] July 1683. Bodleian Library, MS Aubrey 13, f. 242r. Copyright Bodleian Libraries, University of Oxford.

And am very glad to hear tydings of Mr Harriot's papers, (who was a great man in his time,) & hope therefore that they are yet in being. I had formerly heard they had been (at lest some of them) in Mr Hobbes's hands. That they had been, at some other time, in Dr Pell's hands. And that sometime they had been in the hands of the present Bishop of Salisbury. But it is many years since I heard any thing of certainty where they are: and feared they might have perished. I have never read any of his things, but that onely of his Algebra; which hath the good hap to be published by Mr Walter Warner, as a Prodromus to some other of his works; which, at the same time, he gave hopes of publishing; but hath not done it. I concur with you in your opinion, that if that Noble Person (in who's hands, you say, they are) would please to lodge them in the Library of the Royall Society; it would be to them a very acceptable work. But I adde allso, that he would thereby do right to the memory of Mr Harriot; & indeed to the English Nation; who have been very early in the knowledge of these studies; & from whom foreigners have borrowed (to give it no worse a name) very much, without being so kind as to let the World know from whom they had it.[23]

Wallis's official position, then, was the one he shared with most of his contemporaries: that he only knew Harriot through what could be gleaned from the *Artis analyticae praxis* or through reports that had been handed down by his closest associates, notably Walter Warner, Thomas Aylesbury (1576–1657), Robert Hues (1553–1632), and Nathaniel Torporley (1564–1632). Although there is no evidence that Wallis ever met any of these men, he was a good friend of John Pell, who had. And Christopher Wren, for a time as Savilian professor of astronomy Wallis's colleague at Oxford, also had associations with Harriot's circle. As a young man, Wren received a letter of encouragement in his mathematical studies from Thomas Aylesbury, shortly before the latter left England to go into exile in 1649.[24] Moreover, it is unlikely that this letter was the sole communication that took place between the two men.

With Oughtred, things were quite different. Although born 15 years after Harriot, in 1575, he outlived him by 39 years, only dying in 1660. When Wallis was striving to make a name for himself as a relatively young mathematician, in the 1650s, Oughtred was the pre-eminent authority – one might say living authority – of England's mathematical community. He had been a private tutor to a whole generation of mathematical scholars, including Seth Ward, Jonas Moore (1617-79), and Christopher Wren.[25] Wallis had dedicated one of his earliest works to him after his appointment as Savilian professor of geometry at Oxford, in 1649, and the two men stood in correspondence with one another throughout the 1650s.[26] Wallis also

---

23 John Wallis to John Aubrey, 20/[30] July 1683, Oxford, Bodleian Library, MS Aubrey 13, f. 242r-242v. See Shirley, *Thomas Harriot*, 107.

24 Thomas Aylesbury to Christopher Wren, 10/[20] April 1649, in *Parentalia. Or, memoirs of the family of the Wrens*, compiled by Christopher Wren jr and published by Stephen Wren (London: for T. Osborn and R. Dodsley, 1750), 184.

25 See Jacqueline Stedall, 'Ariadne's thread: the life and limes of Oughtred's Clavis', *Annals of science*, 57 (2000), 27–60 (37).

26 John Wallis, *Arithmetica infinitorum, sive nova methodus inquierendi in curvilineorum quadraturam, aliaque difficiliora matheseos problemata*, (Oxford: Leonard Lichfield for Thomas Robinson, 1655).

actively promoted the publication of a fourth edition of Oughtred's *Clavis* in 1667, having already been involved with the third edition of that work in 1652.[27] Compared to Oughtred, Harriot was for most members of England's mathematical community a remote figure from a bygone age (Figure 3.2).

Bringing Harriot to the public

With so little knowledge of Harriot based on the personal experience of key players, it is necessary to ask how he came to have such a powerful influence on the mathematical community in early modern England. To answer this question, we need to take a closer look at Warner, Aylesbury, and Hues, and to consider in some depth the role played by Pell in transmitting their testimony and memory. We know that Walter Warner, an Oxford-educated scholar, was engaged as a servant by Henry Percy (1564–1632), ninth Earl of Northumberland, in 1590, and that he largely took charge of his library and collection of scientific instruments.[28] By the mid-1590s Warner had been joined in Percy's household by Harriot, who in the light of Walter Ralegh's (1552–1618) increasing marginalization at Elizabeth's court had found a new and in many ways ideal patron in the Earl of Northumberland. Scientifically, it was a meeting of minds. Provided with a generous pension and a house on the earl's estate at Syon House, Isleworth, Harriot appears to have had no other obligations than to conduct scientific investigations, particularly on optics, and more specifically the theory of light, an area of interest he evidently shared with his patron.[29]

We know, too, from surviving manuscripts, that both Warner and Robert Hues visited Harriot at Syon House and that the three men conducted optical experiments together. It is probable that Percy himself was involved in such experiments before his arrest and imprisonment in the Tower of London in 1606, but records of experimental work only exist from after that time. Warner, Hues, and Harriot would regularly visit Percy in the Tower in order to discuss their findings with him. Personal discourse was clearly an important part of the circle's activity, and although correspondence also had a place, this was often in order to prepare for and facilitate subsequent meetings. After Hues and Warner had on one occasion used a visit to Syon House to discuss some of his papers on optics, Harriot sent the results to Percy in order that he could consider them before their next meeting in the Tower:

27 Stedall, 'Ariadne's thread', 41–44, 46–50.
28 See John W. Shirley, 'The scientific experiments of Sir Walter Ralegh, the wizard earl, and the Three Magi in the Tower', *Ambix*, 4 (1949), 52–66 (56–57).
29 See Gordon R. Batho, 'Thomas Harriot and the Northumberland household', in Fox (ed.), *Thomas Harriot. An Elizabethan man of science*, 28–47 (37–38).

Sir, When Mr Warner & Mr Hues were last at Syon, it happened that I was perfectting my auntient notes of the doctrin of reflections of bodyes, unto whom I imparted the Magisteryes thereof, to the end to make your Lordship acquainted with them as occasion serued. And least that some perticulars might be mistaken or forgotten, I thought best since to set them downe in writing. whereby allso nowe at times of leasure when your minde is free from matters of greater waight, you may thinke & consider of them, if you please.[30]

Towards the end of his life, Warner provided Pell with a detailed account of experiments to demonstrate Snell's law of refraction, which Harriot, he, and others had conducted 'upon a Table about 2 yards long' at Thomas Aylesbury's house in Windsor. Warner's account makes clear that Harriot especially valued the collective validation of the results of such scientific enterprise: 'he said that they had the judgment of 3 or 4 paire of eyes in this experiment'.[31] Above all, such reports served to keep Harriot's scientific memory alive.

Aylesbury would become an important patron of the mathematical sciences in his own right.[32] He was a close friend of Percy's and probably met Harriot through him. After leaving Oxford, where he had been an undergraduate at Christ Church from 1598 to 1602, Aylesbury became secretary to Charles Howard (1536–1624), lord high admiral. His scientific investigations including astronomical observations were such as he could accommodate alongside his official tasks, but they were immortalized by his friend the poet and divine Richard Corbett (1582–1635), a contemporary of his at Christ Church, on the occasion of the Great Comet of 1618.[33] Aylesbury's surviving letters to Harriot, often written while undertaking official admiralty duties, reveal his devotion to Harriot as a

30 Thomas Harriot to Henry Percy, 13/[23] June 1619, London, British Library, Harley MS 6002, f. 21r-21v; James Orchard Halliwell (ed.), *A collection of letters illustrative of the progress of science in England from the reign of Queen Elizabeth to that of Charles the Second* (London: R. and J. E. Taylor, 1841), 45. See Jon V. Pepper, 'Harriot's manuscript on the theory of impacts', *Annals of science*, 33 (1976), 131–51.

31 London, British Library, Add. MS 4407, f. 183r. See John W. Shirley, 'An early experimental determination of Snell's law', *American journal of physics*, 19 (1951), 507-8.

32 See Anthony Wood, *Athenae Oxonienses. An exact history of all the writers and bishops who have had their education in the University of Oxford: to which are added the Fasti, or Annals of the said university*, ed. Philip Bliss, 5 vols. in 4 (London: for F. C. and J. Rivington et al., 1813-20), vol. 1, col. 305: 'sir Thomas Aylesbury was a learned man, and as a great lover and encourager of learning and learned men, especially of mathematicians (he being one himself) as any man in his time'.

33 See 'A letter sent from Doctor Corbet to Sir Thomas Ailesbury, Decemb. 9. 1618. On the occasion of a blazing star', in *Poems. Written by the Right Reverend Dr. Richard Corbet, late Bishop of Norwich*, 3rd edn (London: J. C. for William Crook, 1672), 54–57. Notable is the line 'Thine own rich Studies, and deep Harriots Mine'.

teacher. And, indeed, Aylesbury never ceases to indicate that he would rather be developing his mathematical and scientific knowledge than undertaking 'wearie journeys' where his only comfort is

> with the remembrance of your kind love and paynes bestowed on your loytering scholar, whose little credit in the way of learning is allwaies underpropped with the name of soe worthie a Maister.[34]

Harriot's relationship with Aylesbury was such that he confidently appointed him as one of four executors of his will, to whom he thereby ultimately gave responsibility for the disposition of his worldly goods. The others were Robert Sidney (1595–1677), second Earl of Leicester, Thomas Buckner (c. 1562-after 1633), a merchant and family friend, and the Welsh landowner and astronomer John Protheroe (or Pretherch) (1582–1624), who had been a close associate of another of Harriot's friends, the politician William Lower (1570–1615).[35] There were manifold connections in play here. Lower, who became a competent mathematician and astronomer in his own right following personal instruction from Harriot, had married the Earl of Northumberland's stepdaughter Penelope Perrot (c. 1590-?) in 1607. In the same year he had made a series of valuable observations of the comet from the Perrot estate in Trefenty, Camarthenshire in order to investigate the nature of its orbit. Protheroe's estate was conveniently close by at Nantyrhebog (or Hawksbrook) and the two men collaborated in their astronomical investigations.[36] There was a good reason also for Sidney being named an executor. He had married the Earl of Northumberland's daughter, Dorothy Percy (c. 1598–1659), albeit secretly, in 1615, and Percy himself had spent some of his retirement at Sidney's residence Penshurst Place in Kent after he had been stripped of public office. Sidney, too, was scientifically interested and in his youth had studied mathematics under Harriot. Finally, Buckner's friendship is reflected in the fact that Harriot spent the last days of his life in his house in Threadneedle Street, London.[37]

Acutely aware of the value of his scientific manuscripts, Harriot entrusted his closest scientific friends with the task of ordering of his papers and seeing them through the press. Explicitly, Harriot's will named Nathaniel Torporley, who had been an undergraduate at Christ Church from 1581 to 1584, and was on paper the most qualified, as overseer of his mathematical

---

34 Thomas Aylesbury to Thomas Harriot, 19/[29] January 1618/19, London, British Library, Add. MS 6789, f. 443r-443v; Halliwell (ed.), *Collection of letters*, 44.

35 Shirley, *Thomas Harriot*, 467.

36 See Francis Jones, 'The squires of Hawksbrook', in *Transactions of the Honourable Society of Cymmrodorion*, yearbook 1937, 339-55 (343-44). See also William Lower to Thomas Harriot, 6/[16] February 1610, London, British Library, Add. MS 6789, f. 427r-429v. There is a seventeenth-century epitaph to the Welsh astronomer in Aberystwyth, National Library of Wales, MS 5390D, f. 181. The text in Latin and English is cited by Jones, p. 355.

37 Shirley, *Thomas Harriot*, 460.

papers.[38] Following his studies at Oxford, Torporley had spent some time in Paris, where he met and consulted with the great Viète.[39] After his return to England, he entered the service of Henry Percy, and went on to produce a number of scientific works, one of which, *Diclides coelometricae*, an exposition of spherical trigonometry, appeared in print in 1602.[40] He was in contact with both Oughtred and Harriot, and no doubt he instructed the two men in some of the basic ideas of Viète's algebraic work.[41] With good reason, Torporley would appear to have been lined up to publish Harriot's mathematical papers or at least ensure that they could contribute in some way to future scholarship. However, Harriot's will contained a significant provision that immediately weakened his position: that in the event of Torporley not understanding some of the notation or writings, he was to consult with Warner or Hues, and if neither of them were able to resolve the issue he was to consult with Protheroe or Aylesbury.[42] In other words, Torporley's role as editor was by no means unconditional, and other members of Harriot's circle proceeded to use this uncertainty to their advantage. Indeed, Torporley was eventually largely excluded by Warner and Aylesbury from further involvement.[43]

By the time of Harriot's death, in 1621, Aylesbury seems to have taken Warner under his wing, accommodating him during the summer months at

---

38 Shirley, *Thomas Harriot*, 469. Rosalind C. H. Tanner suggests, however, that Aylesbury's understanding of Harriot's mathematics was more profound. See her 'Nathaniel Torporley's congestor analyticus and Thomas Harriot's de triangulis laterum rationalium', *Annals of science*, 34 (1977), 393–428 (394-95). See also her 'Nathaniel Torporley and the Harriot manuscripts, *Annals of science*, 25 (1969), 339-49.

39 See Nathanael Torporley to Thomas Harriot, June 1586?, London, British Library, Add. MS. 6788, f. 117r-117v; Jon V. Pepper, 'A letter from Nathaniel Torporley to Thomas Harriot', *British journal for the history of science*, 3 (1967), 285-90; Stedall, *Greate invention of algebra*, 4–5, 17–20.

40 Nathanael Torporley, *Diclides coelometricae seu valuae astronomicae universals* (London: Felix Kingston, 1602).

41 See for example London, British Library, Add. MS 6782, f. 483r-483v: 'A proposition of Vietas delivered by Mr Thorperly but no demonstration'. See also Tanner, 'Nathaniel Torporley's 'congestor analyticus'.

42 Shirley, *Thomas Harriot*, 2. Torporley occasionally wrote to Protheroe on mathematical topics, but there is no evidence that he ever did so on the grounds set out in Harriot's will. See, for example, his paper on triangular numbers for making a table of sines, London, British Library Add. MS 4395, f. 89r-90v; Stedall, *Greate invention of algebra*, 18.

43 It has been suggested that Torporley might himself have re-evaluated his position after discovering Harriot's atomistic theories of light and matter. But this ignores the evidence of substantial disagreement on mathematical matters between the opposing parties charged with taking care of his papers. On Torporley's rejection of Harriot's atomism, and of atomism generally, see his 'Synopsis of the controversie of Atoms', London, British Library, MS Birch 4458, f. 6r-6v. There is a fair copy of this text in another hand at f. 7r-8v. See also Stephen Clucas, 'All the mistery of infinites: mathematics and the atomism of Thomas Harriot', in Sabine Rommevaux (ed.), *Mathématiques et connaissance du monde réel avant Galilée* (Montreuil, 2010), 113-54.

his residence Cranbourne House in Windsor Park. (Aylesbury would be made warden of Cranbourne Chase in 1627.) There is also a considerable amount of evidence indicating that whenever his official duties allowed, Aylesbury would collaborate with his client Warner in his scientific activities. The antiquary Anthony Wood (1632-95) points out that Aylesbury, after Harriot's manuscripts came into his hands, secured an agreement with Henry Percy that his son, Algernon (1602-68), would continue payments to Warner with the aim of ensuring that Harriot's work be published. The two men thus secured controlling influence over future developments. Indeed, evidence suggests that although most of the work on the edition of the *Artis analyticae praxis* was undertaken by Warner, he also had the assistance of Aylesbury:

> The sum of this book coming into the hands of Aylesbury [...] Walt. Warner did undertake to perfect and publish it, conditionally that Algernon eldest son of the said Henry E. of Northumb. would, after his father's death, continue his pension to him during his natural life. Which being granted at the earnest desires and entreaties of Aylesbury made to that lord, Warner took a great deal of pains in it, and at length published it in that sort as we see it now extant.[44]

No mention is made in the published version of *Artis analyticae praxis* as to who was responsible for composing the work.[45] But Wood and his contemporaries at Oxford would almost certainly have been aware of a hand-written note that Robert Payne (1596–1651) had entered on the title page of his personal copy of the book, for this copy was now part of the Savilian Library, the collection of mathematical books initiated by Henry Savile (1549–1622) for the use of his two professors: 'Through the means of Walter Warner, mathematician and philosopher'.[46]

We also have Warner's original draft of the announcement to mathematical scholars, the 'Monitum ad mathematices studiosos', printed at the end of the *Artis analyticae praxis*,[47] and of Torporley's comment questioning the wisdom of Warner's promise – or perhaps threat – that the *Praxis* would be the forerunner of other mathematical works of Harriot to appear in print, the implication being that this would be carried out under his editorship: 'It will doe well in this forme. And I leave it to Mr Warners

---

44 Wood, *Athenae Oxonienses*, vol. 2, 301. On Warner's treatment of Harriot's algebraical work, see Stedall, *Greate invention of algebra*, 20–22.

45 The most thorough investigation of the editorial background to the *Artis analyticae praxis* is by Jacqueline Stedall. See her 'Rob'd of Glories: the posthumous misfortunes of Thomas Harriot and his Algebra', *Archive for history of exact sciences*, 54 (2000), 455-97.

46 Oxford, Bodleian Library, Savile O 9: 'Per Walterum Warnerum, Mathematicum et Philosophum'. Payne had been presented with his copy by Charles Cavendish.

47 See *Artis analyticae praxis*, 180.

discrecion, Whether he think it fit to give this monitum or noe, because he seemed to doubt of it'.[48]

There can be little doubt that Torporley was better qualified to produce the edition of Harriot's work, having been an important interlocutor of Harriot's on questions of algebra, along with another of Harriot's correspondents, William Lower.[49] But, as already mentioned, he seems to have been comprehensively side-lined from the project. All that remained for Torporley to do in the remaining months of his life – he died in April 1632, a year after the publication of the *Artis analyticae praxis* – was to write a damning report on the lacunae and generally muddled composition of the printed work, the 'Corrector analyticus artis posthumae Thomae Harrioti'.[50]

Shortly before Henry Percy's death, in November 1632, Aylesbury wrote to him on behalf of Warner, describing the latter's 'literary labours and paines taken in forming the work and fitting it for the publik view'.[51] This was not the first time Aylesbury had sought to persuade the Earl of Northumberland to recompense Warner for his efforts in bringing about the publication of the *Artis analyticae praxis*, but it was now set out as part of an argument for enabling the future publication, with Warner's assistance, of other mathematical works of Harriot. Aylesbury recognized the uncertainties of times ahead – Algernon would in fact curtail and then stop the payments to Warner – and he evidently sought to mitigate the consequences as far as possible:

> I purpose god willing to set forth other peeces of Master H. wherin by reson of my owne incombrances I must of necessity desire the help of Master W. rather then of any other, Whereto I find him redy enough because it tends to your Lordships service, and may the more freely trouble him, yf he receve some little encoragement from your Lordship towards the repairing of the detriment that lies still upon him by his last imploiment.[52]

48 Nathaniel Torporley, note on 'Ad mathematices studiosos', London, British Library, Add. MS 4395, f. 92r.

49 See, for example, William Lower to Thomas Harriot, 3/[13] April 1611, London, British Library Add. MS 6789, f. 431r; Halliwell (ed.), *Collection of letters*, 41.

50 Nathaniel Torporley, *Corrector analyticus*, London, Lambeth Palace Library, Sion College MS L40.2/E10, ff. 7–12; Halliwell (ed.), *Collection of letters*, 109-16. See Tanner, 'Nathaniel Torporley', 341; Seltman, 'Harriot's algebra', 167; Stedall, *Greate invention of algebra*, 22–24; Lohne, 'Dokumente zur Revalidierung', 191.

51 Thomas Aylesbury to Henry Percy, 5/[15] July 1632, London, British Library, Add. MS 4396, f. 90r.

52 Thomas Aylesbury to Henry Percy, 5/[15] July 1632, London, British Library, Add. MS 4396, f. 90r; Halliwell (ed.), *Collection of letters*, 71. See Shirley, *Thomas Harriot*, 6–7.

As we learn from the letter, Warner's involvement with the publication did not end with editing and composing Harriot's text. It was also he who saw the *Artis analyticae praxis* through the press, a difficult process at the best of times with contemporary mathematical books of a more technical nature, but on this occasion made more difficult by evidently uncooperative printers. By the time the work was completed, Warner was seriously out of pocket, having been supported latterly by his mathematical friends in London. Aylesbury took up his cause, suggesting in a letter to Percy that Warner had carried out the work out of a sense of duty to the duke and implicitly to Harriot as well:

> He looks for no other reward than your Lordships acceptance therof as an honest discharge of his duty. But his long attendance through unexpected difficulties in seeking to get the book freely printed, and after that was undertaken the frivolous delai of the printers and the slow proceding presse, which no intreties of his or myne could remedy, drew him to a gretter expence than his meanes could bere, including both your lordships pension and the arbitrary help of his frends.[53]

Interestingly, Aylesbury wrote a second letter on the same day addressed to a certain Mr T, in which he sought T's approbation and possible backing for the proposition he was making to Percy, but setting out this proposition in rather more modest terms with regard to the long-term financial implications of the intended support for Warner. Naturally, one would be inclined to suspect that this letter was intended for Torporley, the only other figure within Aylesbury's and Percy's circle with such an interest, but he had died some three months earlier, in April 1632, at Syon House.[54] Had Aylesbury failed to hear of the death of the principal overseer of Harriot's mathematical papers? This is somehow hard to imagine, and so one must conclude that he was being deliberately deceptive, perhaps with the intention of conveying the impression that he had been open towards Torporley all along.

---

53 Thomas Aylesbury to Henry Percy, 5/[15] July 1632, London, British Library, Add. MS 4396, f. 90r; Halliwell (ed.), *Collection of letters*, 71. See Shirley, *Thomas Harriot*, 6–7.

54 Torporley's mathematical papers remained at Sion College after his death. See Henry Oldenburg to John Beale, spring 1676, in A. Rupert Hall and Marie Boas Hall (eds.), *Correspondence of Henry Oldenburg*, 13 vols. (Madison, WI and London, 1965-86), vol. 12, 229: 'And there is a MS in Sion College yet unprinted, treating of this argument, wherein there is a Table of figurat numbers, so large, that it is of 3 or 4 sheets of paper. The Author Nathan. Torperly a very learned man, that lived with Vieta till his death, and was his Amanuensis, left this Treatise at Sion College above 40 years agoe'. See also John Collins to Thomas Baker, 19/[29] August 1676, Cambridge University Library, MS Add. 9597/13/5, f.24r-25v; Rigaud, *Correspondence of scientific men*, vol. 2, 4–10, and the summary of Torporley's papers relating to Harriot's mathematics in Jacqueline Stedall, 'Reconstructing Thomas Harriot's treatise on equations', in Robert Fox (ed.), *Thomas Harriot and his world. Mathematics, exploration, and natural philosophy in early modern England* (Aldershot and Burlington, VT, 2012), 53–64 (58–59).

On the trail of Warner's papers

Other cases of contemporary mathematical patronage show just how precarious an existence clients could, and often did, lead. The German-born mathematician Nicholas Mercator (1620?-87), for instance, following an initial stay in London, went to Paris and learnt French explicitly with the aim of improving his chances of future support. After his return to England, he was able to gain employment by the tenth Earl of Northumberland, Algernon Percy, as private tutor to his son, Josceline (1644-70). He was subsequently based at Petworth House in Sussex, where he had the specific brief to teach Josceline mathematics.[55] Mercator's wife, unhappy at her husband's long absence from the family home, hauled him back to London, where he spent more than 20 years as a teacher of mathematics.[56] Mercator's *Logarithmotechnia*, published in 1668, constructed logarithms from first principles and was the first book to publish a function in the form of an infinite series.[57]

Another leading member of England's mathematical community who was always reliant on the financial support of others in order to carry out his scientific activities was John Pell. Most of his mathematical papers also remained unpublished, though few were left in a printable state. Although for a time he had a politically influential patron in William Brereton (1631-80), the Cheshire nobleman was almost as poor as he was. In fact, both Pell and Lord Brereton were at times supported by the lowly accountant and government employee John Collins.[58]

Warner's finances were not any better. In his letter to Percy, Aylesbury points out that Warner's expenses exceeded his means or income made up of his 'Lordships pension and the arbitrary help of his friends',[59] and when Algernon Percy decided to cancel Warner's pension entirely his financial circumstances worsened dramatically. Warner died unmarried and impoverished in March 1643.

---

55 See Samuel Hartlib to John Pell, 3/[13] July 1657, London, British Library, Add. MS 4377, f. 149A-B: 'Hee [sc. Mercator] is to goe next week to Petworth to the E. of Northumberland to teach his son the Mathematiques.'; Wilbur Applebaum, 'A descriptive catalogue of the manuscripts of Nicolaus Mercator, F.R.S. (1620-87), in Sheffield University Library', *Notes and records of the Royal Society of London*, 41 (1986), 27–37.

56 See Joseph Ehrenfried Hofmann, 'Nicholaus Mercator (Kauffman), sein Leben und Wirken, vorzugsweise als Mathematiker', in *Abhandlungen der Akademie der Wissenschaften und der Literatur (Mainz), Mathematisch-naturwissenschaftliche Klasse*, No. 3 (Mainz, 1950), 45–104.

57 Nicholaus Mercator, *Logarithmo-technia, sive methodus construendi logarithmos, nova, accurata, & facilis* (London: William Godbid for Moses Pitt, 1668), 30.

58 See, for example, John Collins to John Beale, 20/[30] August 1672, Cambridge University Library, MS Add. 9597/13/5, f. 83r-85av; Rigaud, *Correspondence of scientific men*, vol. 1, 195–204.

59 Thomas Aylesbury to Henry Percy, 5/[15] July 1632, London, British Library, Add. MS 4396, f. 90r.

Warner's loss of income precluded the possibility of his being able to make any inroads into fulfilling his intention of putting forth 'other peeces' of Harriot. The mathematician's papers were left to collect dust instead. Following Harriot's death, either Aylesbury or Warner carried out their instruction as set out in his will and put his papers into 'a Convenient Truncke' before having this placed in the library at Petworth House, the country estate to which Henry Percy retired following his release from the Tower. They remained there from then onwards,[60] but this archival memory was irretrievably lost for Wallis, Collins, and other members of England's mathematical community, following the deaths of Warner in 1636 and Aylesbury in 1657.[61] It was only after the diplomat Hans Moritz, Count von Brühl (1736–1809) had married the dowager Countess of Egremont Alicia Maria (d. 1794) in July 1767 and discovered Harriot's manuscripts among other household papers at Petworth that their continued existence could be made known to the learned public, notably through Franz Xaver von Zach (1754–1832).[62]

In view of Anthony Wood's description of Aylesbury as 'a singular lover of learning and of the mathematic arts', it was perhaps inevitable that his path would at some stage have crossed with that of John Pell. A skilled mathematician, Pell had corresponded with Henry Briggs (1561–1630) on the topic of logarithms at the early age of seventeen, when completing his undergraduate studies at Trinity College, Cambridge. In the 1630s he had enjoyed close ties to the intelligencer Samuel Hartlib (c.1600–62) and the translator and natural philosopher Theodore Haak (1605-90) in London before moving to the Low Countries, where he taught mathematics for almost ten years, first in Amsterdam and then in Breda, between 1643 and 1652. A detailed record of a problem on proportions proposed to Pell by Aylesbury in January 1638 provides clear proof that the two men knew each other at that time. Although the evidence is only sketchy, it is probable that their acquaintance continued right up to the end of Aylesbury's life, for his final years were spent in exile in Breda and coincide with Pell's time there.

Pell first became aware of Warner through his friend Hartlib who, in 1635, noted in his diary how he had heard from the mathematician Henry Gellibrand (1597–1637) that 'Mr Warner has all Hariots MS and is setting some of them forth'.[63] There is evidence to suggest that Pell worked with Warner on some of the mathematical problems which Harriot had dealt with.

---

60 The main manuscript collections at Petworth House are MS HMC 249 i-v and 241 i-x. The other main collections are London, British Library, Add. MS 6782–6789.

61 See Shirley, *Thomas Harriot*, 7.

62 See Shirley, *Thomas Harriot*, 14–16. This was at least Zach's intention. In 1786 he secured the agreement of the delegates of the University Press at Oxford to publish Harriot's edited works once these were ready, but he failed to deliver. See John J. Roche, 'Harriot, Oxford, and twentieth-century historiography', in Fox (ed.), *Thomas Harriot. An Elizabethan man of science*, 229-45 (239-40).

63 Sheffield University Library, HP 29/3/41 A; Ephemerides 1635.

In a letter to Hartlib, written in January 1639/40, Pell speaks of the defects of the *Artis analyticae praxis* and proposes to overcome these by illustrating and perfecting the examples given in the printed text wherever he considered this necessary. But what is perhaps most remarkable is that Pell at no time faults the man who had prepared the edition, that is to say, Warner, but places the blame for the book's defects squarely with the printer. The account he provides is therefore quite different from that of Torporley:

> To the second Vieta hath taught aequationum resolutionem in numeri somewhat largely but Harriot after him more fully & farre more clearer but the printers faults being many the praecepts too opaque & their grounds not all assured, that booke is not so usefull to learners as it might be, wherefore I have I think so illustrated & perfected that Treatise that it may be easily conceived so perfectly kept in mind that he that hath once understood it, shall never neede to see it againe; but may all wayes be able to write it againe compleately though all copies of it were lost.[64]

During the 1630s, Warner sought the patronage of Charles Cavendish (1620-43) and his brother William (1617-84), both of whom were keen to promote scientific interests across a range of disciplines from mathematics to metallurgy. Most of Warner's correspondence at the time with the Cavendish brothers, as well as with members of their Welbeck Circle such as Robert Payne and Thomas Hobbes, concerns questions relating to optics on which Warner continued to write.[65] Indeed, it was through his work in this area that Warner became more widely known. One of the theorems of his manuscript treatise of refraction was printed later in book six of Marin Mersenne's (1588–1648) *Universae geometriae synopsis* (1644).[66]

But that one theorem was an exception. Just as with the extensive scientific papers left behind by Harriot, which Warner had rashly promised to order and publish, so, too, his own papers were stored away and gathered dust as unpublished manuscripts. Of particular concern to contemporaries was the fate of his so-called canon, his antilogarithmic tables which effectively were to serve as the converse to Briggs's logarithms. This was one of his rare

---

64 John Pell to Samuel Hartlib, 3/[13] January 1639, London, British Library, Add. MS 4419, f. 136r-v.

65 See for example Thomas Hobbes to William Cavendish, 15/25 August 1635, in Noel Malcolm (ed.), *The correspondence of Thomas Hobbes*, 2 vols. (Oxford, 1994), vol. 1, 28–29.

66 See Marin Mersenne, *Universae geometriae, mixtaeque mathematicae synopsis, et bini refractionum demonstratarum tractatus* (Paris: Antoine Bertier, 1644), 548. See also John Collins to James Gregory, 25 March/[4 April] 1671, University of St Andrews Library, ms31009, f. 24r-25v; Herbert Westren Turnbull (ed.), *James Gregory Tercentenary Memorial volume* (London, 1939), 178-81.

mathematical accomplishments. Warner had probably been working on these tables since the beginning of the 1630s. We know that during his first meeting with Pell, in November 1639, Warner informed him of the 'table of Analogicals' he was preparing and showed him the first 10,000 he had calculated. Warner's hope was to extend the table to 100,000 continual proportionals and it seems that already on that occasion the two men discussed this project as a possible enterprise of collaboration. A constant refrain in Charles Cavendish's letters to Pell in 1641 is his desire to hear of progress on the antilogarithmic tables as well as on Pell's own projected work on algebra, building on the advances to this topic made by Viète, Harriot, and others.[67] In view of Warner's advancing years, Cavendish was delighted when, in July 1641, Pell was able to tell him that he was going to assist on completing Warner's canon: 'I am glad you have begun the analogigues, & hope allso that you proceed in your owne analiticall worcke', he writes.[68] A considerable amount of material relating to these tables is to be found among Pell's surviving papers. Some 30 years after Cavendish's note, John Collins describes the collaboration between Warner and Pell in a letter to the Scottish mathematician James Gregory (1638-75):

> One Mr Warner deceased whose Opticks you find mentioned in Mersennus did about 32 yeares since spend above 100^l for ayd, and tooke great paines himselfe, with some assistance from Dr Pell to calculate a table (to 12 places of figures) of 100000 continuall Proportionalls, to witt to find 99999 meane Proportionalls betweene an Unit and 100000, such a large table elegantly writt remaines in the hands of Doctor Thorndyke a Prebend of Westminster.[69]

Following Warner's death, in 1643, his papers came into the possession of his nephew George (seventeenth century), a London merchant. They might not have survived were it not for Nathaniel Tovey (1597–1658), sometime tutor of the poet John Milton (1608-74) at Christ's College, Cambridge, who had married George Warner's sister, Elizabeth, in June 1636.[70]

---

67 See for example Charles Cavendish to John Pell, 26 June/[6 July] 1641, in Noel Malcolm and Jacqueline Stedall (eds.), *John Pell (1611–1685) and his correspondence with Sir Charles Cavendish. The mental world of an early modern mathematician* (Oxford, 2005), 336: 'I desire to knowe if Mr Warners analogicall worck goe on or not.'; Charles Cavendish to John Pell, 26 July/[5 August] 1644, ibid., 354: 'I praye you let me knowe whether Mr: Warners Analogicks be printed'.

68 Charles Cavendish to John Pell, 24 July/[3 August] 1641, in Malcom and Stedall (eds.), *John Pell*, 339; Halliwell (ed.), *Collection of letters*, 73.

69 John Collins to James Gregory, 25 March/[4 April] 1671; University of St Andrews Library, ms31009, f. 24r-25v; Turnbull, *James Gregory*, 178-81.

70 See Gordon Campbell, 'Nathaniel Tovey: Milton's second tutor', *Milton quarterly*, 21 (1987), 81–90.

Letters exchanged between Tovey and his brother-in-law reveal that the two men sought to find a printer willing to take on the task of publishing what Tovey calls 'my Uncle Walters booke', whereby there was a clear division of responsibilities, with Tovey suggesting undertakers, such as the Cambridge printer Roger Daniel (fl. 1627-66) or the London printer Bernard Alsop (or Allsopp) (fl. 1618-53), and Warner having the final say. In one letter to Warner, dated around 1642-43, Tovey writes: 'If you have concluded with a printer there is noe more to be sayde. But if you be yet here, I doe conceive you may have it done by the Cambridge Printer (one Daniel)'.[71] Nothing whatsoever came out of these initiatives for reasons that only become apparent through remarks made later by Pell. Warner was evidently intending to finance the publication, but the collapse of his business put an end to such hopes. As so often happened with plans for more elaborate mathematical books in early modern England, everything hinged on printers and publishers willing to take on the risk. If financial security through subscriptions or otherwise was absent, even the best plans could collapse.[72]

Cavendish, Pell, and Thorndike: resurrecting Warner's project

To the great disappointment of Cavendish, the 'Antilogarithmi Pellio-Warneriani', as the canon came to be known, was never printed,[73] while Pell's *Algebra* did not appear until 1668, being a considerably expanded English translation of the *Teutsche Algebra* of his former Zurich student Johann

---

71 See Nathaniel Tovey to George Warner, c. 1642-43, London, The National Archives, SP 46/83, f. 43r-v. Tovey subsequently adds a remark alluding to the London printer Bernard Alsop (or Allsopp): 'If you be gon already soe farre with a London printer that you cannot goe backe, I desire you to find out Mr Alsop [...] and advise with him for the management of the buisness'. On Daniel, see Cyprian Blagden, 'Early Cambridge printers and the Stationers' Company', *Transactions of the Cambridge Bibliographical Society*, 2 (1957), 275-89.

72 See, for example, Philip Beeley, 'A designe inchoate: Edward Bernard's planned edition of Euclid and its scholarly afterlife in late seventeenth-century Oxford', in Philip Beeley, Yelda Nasifoglu, and Benjamin Wardhaugh (eds.), *Reading mathematics in early modern Europe. Studies in the production, collection, and use of mathematical books* (New York and London, 2020), 192–229.

73 See John Collins to Henry Oldenburg, 30 September/[10 October] 1676, in Hall and Boas Hall (eds.), *Correspondence of Henry Oldenburg*, vol. 13, 83–89: 'Betweene the yeares 1630 and 1640, Dr Pell and one Mr Warner deceased mentioned in Mersennus, agreed to make a table of Antilogarithmes, which were called Antilogarithmi Pellio Warneriani, and accordingly such a table was computed and left in the hands of Dr Thorndyke deceased, and cost Mr Warner above 40 Crownes the doing'. See also Charles Cavendish to John Pell, 16/26 August 1644, London, British Library. Add. MS 4278, f. 153; Halliwell (ed.), *Collection of letters*, 83: 'I am sorie Mr Warners analogicks are not printed, but I yet hope they maye, as also other worckes of that excellent olde man'.

Heinrich Rahn (1622-76).[74] In response to Cavendish's latest exhortation, sent from Hamburg in the summer of 1644, that he proceed with his 'intended worcke of Analiticks' coupled with a request for news on the printing of the Analogicks,[75] Pell's reply was joyless. At the time, Pell was in Amsterdam and as he informs Cavendish, when leaving England he had, with a view to their safety, left his papers behind. That might have seemed a prudent thing to do at the time, but now with 'the disasters of the whole Kingdome', that is to say, the extending military conflicts of the Civil Wars, he was beginning to have second thoughts. He was reconsidering especially his earlier disinclination to publish. The comments he makes thereby were becoming increasingly relevant for the unpublished papers of Harriot and Warner, too:

> I have thought nothing elaborate enough to be printed, till it were so complete, that no man could better it and did therefore so long keepe my name out of the presse: But now I begin to count nothing safe enough till it be printed and therefore I have almost resolved to secure my thoughts, not by burying my papers in England, nor by fetching them hither but by publishing the same notions heere, that I have committed to paper there.[76]

As to Warner's Analogiques and the rest of his papers, Pell had even more depressing news to report, for, as he writes, one of Warner's relatives, 'a London merchant', into whose hands they had come following his death had, together with his partner, become bankrupt. Pell was referring of course to George Warner. All indeed seemed lost. But here, at least, Pell's wonderful black humour, indeed, mathematical humour, was able to shine through:

> I am not a little afraid that all Mr Warners papers, and no small share of my labours therein, are seazed upon and most unmathematically divided betweene the Sequestrators and Creditors, who (being not able to ballance the account when there appear so many numbers, and much troubled at the sight of so many crosses and circles in the superstitious Algebra and that blacke art of Geometry) Will no doubt

74 *An introduction to algebra, translated out of the High-Dutch into English, by Thomas Brancker. M.A. Much altered and augmented by D[r] P[ell]. Also a table of odd numbers less than one hundred thousand, shewing those that are incomposit,* and resolving the rest into their factors or coefficients, London: W. G. for Moses Pitt 1668.

75 Charles Cavendish to John Pell, 26 July/[5 August] 1644, London, British Library, Add. MS 4278, f. 149r; Halliwell (ed.), *Collection of letters*, 78. Fearing that this letter might not have reached Pell's hands, Cavendish wrote again nearly two weeks later. See Charles Cavendish to John Pell, 8/18 August 1644, London, British Library, Add. MS 4278, f. 151; Halliwell (ed.), *Collection of letters*, 79: 'I desire to knowe if Mr Warner's Analogicks be printed, and if there be any newe books of Analiticks, but I expect no greate advancement of Analitickes but by yourself, therefore I beseech you proceede in your intended worcke'.

76 John Pell to Charles Cavendish, 10/20 August 1644, in Malcolm and Stedall (eds.), *John Pell*, 357-59, esp. 359; Halliwell (ed.), *Collection of letters*, 79–81.

determine once in their lives to become Figure-casters and so vote them all to be throwen into the fire; if some good body do not reprive them for pye-bottoms etc for which purposes you know Analogicall numbers are incomparably apt, if they be accurately calculated.[77]

At that point the trail goes cold. But in December 1652, the theologian and biblical scholar Herbert Thorndike (1597–1672) writes to Pell after Nathaniel Tovey had placed Warner's papers at his disposal. We must conclude, therefore, that following George Warner's bankruptcy his uncle's papers had passed to Tovey.[78] Thorndike gives Pell a brief breakdown of the what is now in his hands, including a number of copies of the antilogarithmic tables in various stages of completion, and asks his assistance in finally resolving what was to be published. In order to avoid any question of dispute, Pell was to provide a certificate of receipt:

Mr Tovey is gone out of towne and hath left mee to dispose of Mr Warners papers, which I would have had done himself. I have therefore sent you whatsoever I can conceive to concerne the canon: being, 1) The canon itselfe from 1 to 100,000. 2) A collection of papers sowed together, concerning (I suppose) the construction and use of it, intituled on the front Tabularia. 3) Papers of Interest and the questions of it, sowed together. 4) A peece by itself of about halfe a quire, beginning with 'Any ratio being given'. 5) The canon from 1 to 10,000. 6) A foule copy of the same, in which are bound up other loose papers concerning the subject. 7) Foule papers in nine bundles, which seeme to be the first copy of the large canon. And my request to you is, first that you will take your own time to peruse them, in order to a resolution of publishing them, which, upon perusing them, I hope you will declare: and then in consideration of common casualties, and the uncertainty of my continuing where you are, that you will certifie mee of the receipt of the particulares.[79]

It is perhaps surprising that a biblical scholar should be concerned with such mathematical matters. But Thorndike, who after the Restoration would become prebendary of Westminster Abbey, was closely associated with another promoter of mathematics, Sir Justinian Isham (1611-75) of Lamport Hall, Northamptonshire, whose circle of scientifically minded friends included Seth Ward and Samuel Hartlib, both of whom we have met before.[80] Thorndike

---

77 Ibid.

78 See Stedall, 'Rob'd of Glories', 477.

79 Herbert Thorndike to John Pell, 23 December 1652/[2 January 1653], London, British Library, Add. MS 4279, f. 261r; Halliwell (ed.), *Collection of letters*, 94.

80 See Norman Marlow (ed. and trans.), *The diary of Thomas Isham of Lamport (1658-81). Kept by him in Latin from 1671 to 1673 at his father's command* (Farnborough, 1971), 133. Hartlib suggests that Warner's mathematical manuscripts went for a time to Isham, and

had been a contemporary of Pell's at Cambridge, although he had preceded him at Trinity College by over ten years, and it is possible that the two men had met there already.[81] Incidentally, Thorndike's strong leaning towards the Roman Catholic church on doctrinal issues led to his being championed by members of the Oxford movement in the nineteenth century, including John Henry Newman (1801-90), a fellow of Oriel College. Thorndike approached Pell again in June 1655, conveying to him the disappointment of Tovey and his friends at Pell's failure to publish Warner's canon and now requesting that all the papers be returned.[82] Unfortunately, Pell was already in Switzerland at the time, engaged in a diplomatic mission for Cromwell, and so passed on the task to his wife Ithamar back in London: 'I hope you remember that the box with the broken cover in my study holds Mr Warner's papers in it. I would have you deliver it, as it is, to Mr Thorndike', Pell writes from Zurich on 14 July 1655.[83] In a separate letter, he explains to Thorndike the reasons for not returning the papers before his departure and the need for a person of trust, such as Hartlib or someone else known to his wife to accompany Thorndike when he went to collect them:

> The papers, for which you have written, I did not send you, before I went from home, because I had some reasons to expect a speedy returne. But I gave my wife charge of the box wherein they were, that she should let nothing be taken out of it, but keepe it safe till I came back or you sent for it. And therefore, though she doe not know you; yet she will make no difficulty to deliver it to your selfe, if you be accompanied with Mr Hartlib or any other whom she knowes.[84]

Twelve years later, in December 1667, the mathematical intelligencer John Collins borrowed the various manuscript sheets and bundles from Thorndike and listed these somewhat haphazardly on a separate piece of paper as 'An Inventorie of the Papers of Mr Warner'. The list contains twenty-three different items, ranging from tracts on coins and ingots to a fair copy of Warner's canon of 100,000 antilogarithms. The writings on

---

that Charles Thynne showed them to Pell, but there is no further evidence for this: Sheffield University Library, HP 28/2/29 A; Ephemerides 1653.

81 See Malcolm and Stedall (eds.), *John Pell*, 17–18.

82 Herbert Thorndike to John Pell, 21 June/[1 July] 1655, London, British Library, Add. MS 4364, f. 167r-167v: 'It was very much desired by Mr Tovy & his friends for the remembrance of his Uncle, & the good of learning, to see Mr Warners Canon published by you, as the person that could give most lustre to it […] And hath therefore desired mee, from whose hands you received the Papers concerning that worke, to move you to send hither, where we understand they remaine, that they may be redelivered to my hands'.

83 John Pell to Ithamar Pell, 14/[24] July 1655, in Robert Vaughan (ed.), *The Protectorate of Oliver Cromwell, and the state of Europe during the early part of the reign of Louis XIV*, 2 vols. (London: Henry Colborn, 1838), vol. 2, 398-99.

84 John Pell to Herbert Thorndike, 14/24 July 1655, British Library, Add. MS 4364, f. 168r-v.

metals and coinage were produced at the behest of his patron Aylesbury, who from 1635 onwards had served as Master of the Mint:

An Inventorie of the Papers of Mr Warner

1   A Tract of Exchanges in folio cont 11 leaves anglice
    Varronis Sententia de Tympanis illustrata, 3 foliis
2   A treatise of Coines, 3 Another of the same
3   A tract about Ingotts, Another of the same 5
4   Opus Saturninum
    A Bundle containing 30 Papers Intituled Opus Joviale
    A small Bundle intituled Observationes Westmonasterienses
    A Bundle intituled Monetary
    A Bundle intituled Generall rules of Warre and Fortification observed by the Experiences of Richard Harbord
    Six tracts sewed together intituled Tabularia
    A faire Copy of a Canon of 100000 Logarithmes
    Canones analogici Originalis
    Schedae Miscellaneae
    A Bundle intituled Analogity Analyticks
    De Monetarum homonimicarum æquivalentia
    De Resectione Spatii.
    A treatise sic incipiens Any ratio being given
    A treatise thus beginning, Of that columne [check that]
    A bundle de Refractione Definitiones
    A bundle intituled Mr Protheroe
    A bundle intituled Sir William Beecher[85]

This was evidently an indefinite loan, for on his hand-written inventory exchanged with Thorndike at the time Collins only undertakes to restore the papers when requested to do so.[86] It is likely that he still had them in his possession in the following March, when he sought to persuade his friend James Gregory (1638-75), professor of mathematics at St Andrews, of the potential usefulness of Warner's propositions for determining the quadrature of the hyperbola. Unfortunately, the mathematical intelligencer met with a less than enthusiastic response from the Scottish mathematician.

Pell's desire to keep his name out of the press did not go unnoticed. On one occasion, Collins wrote of Pell, 'to incite him to publish any thing seems to be

---

85 John Collins, 'An Inventorie of the Papers of Mr Warner', London, British Library, Add. MS 4394, f. 105r; Halliwell (ed.), *Collection of letters*, 95. Sir William Beecher (1580–1651), sometime Clerk of the Privy Council, enjoyed close ties with Thomas Aylesbury.
86 Ibid.: 'December 14th, 1667. Received the abovesaid Papers from Dr Thorndyke, which I promise to restore upon Demand – John Collins.'

as vain an endeavour, as to think of grasping the Italian Alps in order to their removal'.[87] Cavendish made similar remarks, having witnessed how the planned collaboration on his canon could not be brought to a successful conclusion.[88] It was also true of work in which Pell played the principal role. When finally his *Introduction to algebra* came out in 1668, after many years of production, Pell hid his name behind his initials, while giving far more prominence to his collaborator Thomas Brancker (1633-76). But there was a widespread reluctance among English mathematicians to publish their work throughout the seventeenth century, and the disinclination of printers and publishers to take on mathematical work of a more theoretical nature did not make things easier. Warner's failure to fulfil his promise, set out at the end of *Artis analyticae praxis*, to publish those remaining tracts of Harriot which were in an advanced stage of completion was a mixed blessing. It meant that other works of Harriot were spared the treatment of that book and the likely loss of associated manuscripts in the process. But it effectively deprived contemporary audiences of further concrete insights into the richness of Harriot's achievements particularly in the fields of algebra, combinatorics, optics, and mechanics – achievements which have only been revealed in the past few years thanks to the efforts of the late Jackie Stedall and Matthias Schemmel, and more recently Robert Goulding.[89]

Wallis, Collins, and Aubrey: the unfulfilled promise of Harriot's papers

There were, as we know, efforts to make up lost ground already during the second half of the seventeenth century. The Royal Society commissioned investigations into the whereabouts of Harriot's papers in 1662 and 1663, but these produced little more than confusion as to their fate since the passing of Harriot's chief executor Thomas Aylesbury. As mentioned earlier, it was suspected that through the marriage of Aylesbury's daughter Frances (1617-67) to Edward Hyde (1609-74), first Earl of Clarendon, they had come into the possession of the lord chancellor. Matthew Wren (1629-72), Hyde's secretary, reported in November 1662 that the lord chancellor was prepared to communicate those Harriot papers in his possession to the Society, but which papers these were was not revealed. Nor were they ever forthcoming.[90]

87 John Collins to John Beale, 20/[30] August 1672, Cambridge University Library, MS Add. 9597/13/5, f. 83r-85av; Rigaud (ed.), *Correspondence of scientific men*, vol. 1, 195–205 (196).

88 See, for example, Charles Cavendish to John Pell, 25 July/[5 August] 1644, in Malcolm and Stedall (eds.), *John Pell*, 354. See also p. 109.

89 See the online platform hosted by the Max-Planck-Gesellschaft für Wissenschaftsgeschichte: https://echo.mpiwg-berlin.mpg.de/content/scientific_revolution/harriot/harriot_manuscripts.

90 See the minutes of the meeting of the Society on 19 November 1662 (old style), in Birch (ed.), *History of the Royal Society*, vol. 1, 126: 'Mr. Matthew Wren acquainted the society, that the lord chancellor, upon the intimation of their desire, had expressed his readiness to

Subsequently it was rumoured that these papers had passed to his son Henry Hyde (1638–1709), Lord Cornbury, who remained in England after the lord chancellor's downfall and passage into exile, and eventually became the second earl. Since the eponymous family estate was near to Oxford this would have placed them tantalizingly close to Wallis. Alas, nothing actually seems to have been there in the first place.

In December 1669, Collins informed the Society that 'many papers' of Harriot were in the hands of Richard Vaughan (c. 1600–86), son of the earl of Carberry.[91] Vaughan was appointed president of the Council of Wales and the Marches after the Restoration and had a reputation as a literary patron, but his link to Harriot was through the mathematician's Welsh executor Protheroe. This line of archival descent was no less complicated than that from Aylesbury, and it came about through Protheroe's marriage to Elinor Vaughan (16th/17th century), sister of the first Earl of Carberry, John Vaughan (1574/5–1634).[92] Significantly, Richard Vaughan was succeeded as third earl by John Vaughan (1640–1713), sometime governor of Jamaica, who served as president of the Royal Society from 1686 to1689. Again, the intelligencer John Collins is our main source of information. Writing to the diplomat Francis Vernon (1637?-77) in Paris in December 1671, Collins points out that although most of Harriot's papers following his death went to Aylesbury, some pieces on algebra went to Protheroe and some further sheets concerning the laws of motion to the president of the Royal Society, William Brouncker (1620-84).[93] But this was only the beginning of a long and sometimes tortuous archival journey reflecting the vagaries of family inheritance that so often in the seventeenth century left the survival of England's scientific heritage hanging in the balance:

> His papers fell into the hands of Sir Thomas Aylesbury, who was father to the later Lord Chancellor's lady, by which means they fell into the Lord Chancellor's hands, to whom application was made by members of the Royal Society to obtain them: his lordship (then at the height of his dignity and employments) gave order for a search to be made, and in

communicate to them several papers of Mr. Harriot, which he had in his custody; and that her would give Mr. Wren access to his trunks for them'. See also pp. 120, 309.

91 See the minutes of the meeting of the Society on 3 December 1669 (old style), in Birch (ed.), *History of the Royal Society*, vol. 2, 410: 'Mr. Collins mentioned, that he had been informed, that many papers of the famous mathematician Mr. Thomas Harriot were in the hands of the son of the Earl of Cherbury. Upon which Mr. Oldenburg said, that he would endeavour to procure a sight and transcript of them, if they were in those hands.' The minutes clearly mistake Cherbury for Carberry here. See also Shirley, *Thomas Harriot*, 9.

92 Jones, 'Squires of Hawksbrook', 345.

93 Lohne points out that the complete tract has been preserved at Petworth House. See Johannes Lohne, 'Essays on Thomas Harriot. I: Billiard balls and laws of collision', in *Archive for history of exact sciences*, 20 (1979), 189–229 (190, 214-15).

result the answer was, they could not be found. I am afraid the search was but perfunctory, and that, if his lordship (now at leisure) were solicited for them, he might write to his son the Lord Cornbury to make a diligent search for them. One Mr Protheroe, in Wales, was executor to Mr Harriot, and from him the Lord Vaughan, the Earl of Carberry's son, received more than a quire of Mr Harriots Analytics. The Lord Brouncker has about two sheets of Harriot de Motu et Collisione Corporum, and more of his I know not of: there is nothing of Harriots extant but that piece which Mons. Garibal hath.[94]

Wallis remained unsure of the papers' fate throughout the 1670s and 1680s, having heard that they had been in the hands of various contemporaries with whom he was well acquainted, such as Pell and Hobbes. He evidently knew Matthew Wren fairly well and was confident that nothing remained in the possession of the Hyde family. Wallis also had very good ties to Collins and could be sure through his exchanges with him that other possible locations of Harriot's mathematical papers had been exhausted. Thus, we find that numerous fruitless enquiries by John Wallis eventually led the Savilian professor of geometry at Oxford to the desperate – and thankfully wrong – conclusion that little had survived. Indeed, writing to Aubrey in 1684, Wallis suspects that the only surviving papers might be those in the possession of the legendary head master of Westminster School, Richard Busby (1606-95), after a recent search of the second Earl of Clarendon's Oxfordshire residence at Cornbury, that Wallis had delegated to Matthew Wren, had turned up nothing:

Since I saw you, I have made application to the Earl of Clarendon (by a person well acquainted with him,) concerning those papers of Mr Harriot which were supposed to be in his hands. He sayth, He hath made search for them at Cornbury; but doth assure us he hath them not. So that, I guess, here are no other of them to be found, than those in Dr Busby's hands; who hath divers papers of Mr Walter Warner's (the publisher of what is extant of Harriots) amongst which may be what remains of Harriot.[95]

---

94 John Collins to Francis Vernon, mid-December 1671, Cambridge, Cambridge University Library, MS Add. 9597/13/5, f. 68r-69v; Rigaud (ed.), *Correspondence of scientific men,* vol. 1, 151-56. There is no indication of a person under the name of Garibal in contemporary literature.
95 John Wallis to John Aubrey, 8/[18] March 1683/84, Oxford, Bodleian Library, MS Aubrey 13, f. 243r-243v. The following year, in notes he prepared for his brief life of Harriot, Aubrey suggests on the basis of testimony from Pell that the papers were now with Richard Busby. See Kate Bennett (ed.), *Brief lives with an apparatus for the Lives of our English mathematical writers,* 2 vols. (Oxford, 2015), vol. 1, 109-10: 'Dr Pell tells me, that he finds amongst his papers (which are now in Dr Busby's hands), An Alphabet that he had contrived for the American Language: like Devills'.

In the absence of Harriot's papers themselves, it was only possible for Wallis to refer Aubrey to John Pell, who could at least provide an account of what they contained, for he had seen and studied them with his own eyes. 'Dr Pell, (who hath seen & perused them,) can give you a better account of them', he writes.[96]

The tantalizing promise, the mystique, remained. And Pell was the chief witness to what the papers contained. 'As to Harriot', writes John Collins to the Francis Vernon in Paris in early 1672, 'he was so learned, saith Dr Pell, that had he published all he knew in algebra, he would have left little of the chief mysteries of that art unhandled'.[97]

But the final word should remain with a poet, Edward Sherburne (1616–1702), a close friend and colleague of the practical mathematician Jonas Moore (1617–79), a former pupil of Oughtred's. In his English translation of *The Sphere of Marcus Manilius* (1675), Sherburne has a section devoted to contemporary mathematicians and astronomers. And there, after talking of the Hamburg mathematician Joachim Jungius (1587–1657), many of whose works remained unprinted at his death 'for want of due Encouragement', he says of Harriot:

> On the like Reasons we may conceive we want the many learned Algebraical Works of our famous Countryman Mr. Thomas Harriot, (and of Mr. Warner, into whose Hands they fell) who is esteemed by some of the most knowing Persons alive to have been much Superiour to all that ever writ; and, that equivalent to what of his might have been forty or fifty years since known, is not readily to be expected.[98]

Conclusion

Against the backdrop of rapid advances in the mathematical sciences across continental Europe in the first half of the seventeenth century, Thomas Harriot and to a lesser extent William Oughtred came to epitomize English designs to stand as equals, if not achieve pre-eminence, alongside Italy, France, Germany, and the Low Countries. It was crucially important thereby that these two mathematicians were able to make their mark, albeit in different ways, in the field of algebra, for following on from Viète, Cardano, Tartaglia, and others the study of what came to be known as 'analysis speciosa' had quickly developed into the focal point of contemporary mathematical interest in England, from practical milieus in London and elsewhere through to the universities of Oxford and Cambridge.

---

96 John Wallis to John Aubrey, 8/[18] March 1683/4, Oxford, Bodleian Library, MS Aubrey 13, f. 243r-243v.

97 John Collins to Francis Vernon, mid-January 1671/72, Cambridge, Cambridge University Library, MS Add. 9597/13/5, f. 68r-69v; Rigaud (ed.), *Correspondence of scientific men*, vol. 1, 151-56.

98 Edward Sherburne, *The Sphere of Marcus Manilius made an English poem. With annotations and an astronomical appendix* (London: for Nathanael Brooke, 1675), 118.

Harriot's reputation as algebraist rested largely on his posthumously published *Artis analyticae praxis*, a work that despite its many introduced imperfections enabled something of the author's modernity and genius to shine through.[99] Additionally, reports from former members of his circle such as Aylesbury and Warner allowed the promise of what might be contained in his other surviving scientific papers to take on legendary character within England's mathematical community. Over time, these reports were increasingly overshadowed by fears that apart from some scattered remains the papers might not have survived. And thus, Harriot came also to epitomize another side of the scientific culture of early modern England, the loss of intellectual heritage due to either unwillingness or inability to publish or the lack of a suitable national institution to house and safeguard unpublished material.

There were personal failings, too. The initial fate of Harriot's papers was sealed by his attempt to assign all the members of his immediate scientific circle significant roles in their preservation and publication. Although the eminently qualified Torporley was given oversight of the papers, Harriot appears to have been concerned by his intellectual independence, his individual skill as a mathematician, leading him to give others, especially Aylesbury and Warner, decisive powers of intervention. Harriot's one mathematical publication was consequently the result of their inadequate editorial efforts, with Torporley effectively consigned to the role of spectator.

Warner, like Torporley, had mathematical ambitions of his own, but was less successful in bringing his work to a conclusion or in finding adequate financial support to carry it out. He, too, came to reflect the uncertainties of scientific life in early modern England, his papers enduring an odyssey of ownership and fears of irrevocable loss, similar to what was imagined in the case of Harriot. Well into the second half of the seventeenth-century discussions on the fate and whereabouts of the surviving papers of figures such as Harriot and Warner gave the country's mathematical community cause to reflect on its international standing and its intellectual values. Not least in this way, Harriot posthumously came to play a considerable role in forging that community's own sense of identity.

Note

All dates are given Old Style, that is to say, according to the Julian calendar as used in England used in England until 1752, with the start of the year from Lady Day (25 March). For ease of reference against standard editions, dates of correspondence are additionally given New Style.

The author would like to thank the Bodleian Libraries, University of Oxford, for granting permission to publish copyright material in their possession.

---

99 See Jon V. Pepper, 'Thomas Harriot and the great mathematical tradition', in Fox (ed.), *Thomas Harriot and his world*, 11–26 (13–14).

4   Thomas Harriot: The World's First Ethnographer?

Mark Horton

The very first Harriot lecture was delivered in 1990 by the scholar and historian of early British colonization of the New World, David Beers Quinn. Quinn so dominated the field, with his many publications, that for many years, historians doubted that any more could be said of Ralegh's Virginia ventures and the involvement of Thomas Harriot. In his contribution to Harriot studies, Quinn published three papers: his 1990 lecture, a paper in the Shirley volume on Harriot, and a short paper on Harriot and Ralegh's 1602 voyage.[1] Since 1990, few of the Harriot lectures themselves have addressed Harriot's New World activities in the 1580s; those that have done so have been chiefly concerned with the navigational aspects of his remarkable mathematical mind.

The justification for this paper is that 30 years on, there is now new information, mostly archaeological, that was not available to Quinn at the time. I write as an historical archaeologist who has been working over the last ten years on the remains of the Native American sites of the Outer Banks that Harriot described in such detail, as well as discovering remnants of the Elizabethan colonists who were abandoned there in 1587 and who subsequently disappeared from history. These investigations have provided us with much new data about the Algonquian way of life in the sixteenth century, the topography of the Outer Banks, and how it has changed since Harriot's time, as well as new discoveries relating to the Elizabethan activities.

The 'Roanoke voyages', as they were termed by Quinn, have attracted a large historical literature since he was writing, and made available the many

1 D. B. Quinn, 'Thomas Harriot and the problem of America', in R. Fox (ed.), *Thomas Harriot. An Elizabethan man of science* (Aldershot and Burlington, VT, 2000), 9–27; 'Thomas Harriot and the New World', in John W. Shirley (ed.), *Thomas Harriot. Renaissance scientist* (Oxford, 1974), 36–53; 'Thomas Harriot and the Virginia voyages of 1602', *William and Mary quarterly*, 27 (1970), 268–81.

DOI: 10.4324/9781003096580-5

documents in English transcription.[2] Popular fascination around the 1587 settlement (which did not include Harriot), the so-called 'lost colony', has continued since Paul Green's 1937 play, which is performed every summer in the open air on Roanoke island.[3] The voyages and the fate of the colonists have become a foundation story of English North America, and whether the colonists survived and were assimilated into native American culture, or were killed, or simply died of starvation. The mystery has generated conflicting views as well as forgeries about the origins of English America, and especially the fate of Virginia Dare, John White's granddaughter, and the first English birth in North America.

This paper is sub-titled with the bold claim 'The world's first ethnographer?'. Ethnography as a discipline is often characterized as emerging in Germany in the eighteenth century, as *Völker-Beschreibung,* and was first practiced in what has become known as the Great Northern Expedition to explore the Arctic regions in 1733–43.[4] Others suggest that it emerged in 1914, with Malinowski's study of the Trobriand islanders, which at its base has *participant observation* of indigenous cultures or in the nineteenth century with the work of E.B.Tylor and Lewis Morgan.[5]

So, what justifies our claim that Harriot was actually pioneering the method of 'ethnography' during the great period of Elizabethan science in the later sixteenth century? Two methods of later ethnographers are already present in Harriot's work. The first is linguistics. Harriot took considerable care to learn and record the Algonquian dialects that he was going to encounter on the Outer Banks. Documentary evidence of his linguistic notations survives in papers at Westminster School that were unknown to Quinn in his major publications.[6] It is clear that Harriot mastered the Algonquian language and was able to freely communicate with the native Americans.

Such linguistic skills were however not enough to qualify as an ethnographer. Harriot's only publication in his lifetime was *A briefe and true report of the new found land of Virginia.* An illustrated printed edition was

2 D. B. Quinn, *The Roanoke voyages, 1584–1590,* 2 vols. continuously paginated (London, 1955); *Set fair for Roanoke. Voyages and colonies 1584–1606* (Chapel Hill, 1985); Phil Jones, *Ralegh's pirate colony in America* (Stroud, 2001); Michael Leroy Oberg, *The head in Edward Nugent's hand. Roanoke's forgotten Indians* (Philadelphia, 2008); James Horn, *A kingdom strange. A brief and tragic history of the lost colony of Roanoke* (New York, 2010); Andrew Lawler, *The secret token* (New York, 2018); Scott Dawson, *The lost colony and Hatteras Island* (Charleston, 2020).

3 Laurence G. Avery, *The lost colony. A symphonic drama of American history by Paul Green* (Chapel Hill, 2001).

4 H. F. Vermeulen, 'Gerhard Friedrich Müller and the genesis of ethnography in Siberia', *Etnografia,* 1 (2018), 40–63. doi:10.31250/2618-8600-2018-1-40-63.

5 B. Malinowski, *Argonauts of the Western Pacific. An account of native enterprise and adventure in the archipelagoes of Melanesian New Guinea* (London, 1922); E. B. Tylor, *Primitive culture* (London, 1871).

6 Westminster School Archives, Personal papers of Thomas Harriot, GB 2014 WS-05-HAR.

translated into four languages, with engravings supplied by Theodor de Bry and extended captions, written by Harriot, in 1590. These engravings were based on John White's famous watercolours, drawn in the field alongside Harriot's explorations in 1586.[7] Taken together, Harriot's text and White's drawings provide one of the most remarkable and sympathetic visualizations of native American culture at the point of contact.

Harriot was also able to compile detailed maps of the region that he was visiting. The maps were not simple estimations; they were based upon mathematical principles that must have involved advanced surveying techniques. The Harriot-White partnership enabled them to be produced to a high level of detail and artistic quality, including not only topography, but also the location of settlements, and the incorporation of Algonquian clan and place names. This cartographic record greatly adds to the ethnographic information in providing a sixteenth-century spatial geography of the Outer Banks.

Harriot and the Roanoke voyages

With the death of Sir Humphrey Gilbert, the formal rights of North American exploration passed to Walter Ralegh, with a patent dated 16 March 1584. Ralegh had already drawn around him a number of scientists at Durham House, during the winter of 1583–84, including the young Thomas Harriot, who delivered instruction on the use of navigational instruments and the construction of sea-charts. The information that he imparted was written up in the now lost *Arcticon* and was put to immediate use in the first Roanoke voyage, the reconnaissance by Philip Amadas and Arthur Barlowe.[8] This expedition set sail on 27 April 1584 and arrived off the Outer Banks, landing on 13 July at an unknown spot close to modern Hatteras Island. Members of the expedition later met with local native Algonquians, most probably on Roanoke Island. Amadas and Barlowe's stay on the Outer Banks lasted around six weeks, and they returned with two Algonquians, Manteo and Wanchese.

There has been some debate as to whether Harriot was present on this early voyage. Quinn suggests that John White was, since he later describes his final voyage of 1590 as his fifth to the New World, even though White is not specifically mentioned in the ten names on the voyage.[9] The Barlowe-Amadas narrative and careful observations about the customs of the local inhabitants suggest input from Harriot;[10] the White-Harriot partnership may have started at this point. Harriot was able to understand Algonquian

7 Paul Hulton, *America 1585. The complete drawings of John White.* (London and Chapel Hill, 1984); Kim Sloan, *A new world. England's first view of America* (London, 2007).

8 Quinn, 'Thomas Harriot and the New World', 38–39.

9 i.e., 1584, 1585–86, 1587, 1588, and 1590.

10 Quinn, *Roanoke voyages*, 91–115.

and include promotional information for the future plans in late November 1584, only two months after the arrival of Manteo and Wanchese in England. If he had been on the voyage, this would have provided a longer period for mastery of the language (and probably for Manteo and Wanchese to acquire some English). But given Harriot's linguistic ability, two months may well have been sufficient. One remarkable survival is a page of phonetic transliteration of the Algonquian in the archives of Westminster School that Harriot compiled in the winter of 1584–85. It is likely that he spent much time with the two Virginians, preparing for the 1585 voyage, and he may have visited the West Country. A plaster overmantel at Greenway House (Devon), owned by the Gilberts, where Ralegh is known to have visited at this time, shows two exotic figures smoking stylized pipes, possibly a folk memory of their visit.

Harriot (and White) left with Sir Richard Grenville in April 1585, first reaching the West Indies (where they collected horses for the new colony in Puerto Rico) then sailing up the American coast. It is likely that Harriot and White were on board the flagship *Tyger*, along with Manteo and Wanchese; while Harriot was employed for his navigational skills, White compiled plans of a fortified encampment and an entrenchment for collecting salt in Puerto Rico as well as drawing its natural history. The forts show exceptional military detail, suggesting that Harriot may have been involved in their layout.[11] The fort built on Roanoke Island could have followed the Puerto Rico plan, adjacent to a creek and open along the shoreline.

Their arrival on the Outer Banks was a disaster. On 26 June, an inlet was located by the island of Wococon (the modern Ocracoke Island); the smaller ships grounded trying to enter the sound but were successfully refloated. On 29 June the *Tyger* tried to pass through the inlet but also ran aground and was beached, with the loss of stores and many of the animals that had been purchased in Puerto Rico, although all the crew and passengers survived. From the early base on Ocracoke, Grenville set out, on 11 July, in a 'tilt boat', a pinnace, and two other boats to cross the very shallow Pamlico sound, with Harriot and White on board. White drew the Indian villages of Pomeioc and Secotan as well as their inhabitants. A rudimentary sketch-map[12] of this area of the Outer Banks also survives, which was sent back to England in September, suggesting that serious survey had yet to commence.

11 Kim Sloan, *New world*, 100–2. These two plans are all that survives of fort details from White's drawings. The encampment is approximately 100 paces across, around 80 m. As this was erected in a few days, any similar fortification on the Outer Banks is likely to have been at least this size, though in all probability larger. The entrenchment, though not scaled, may have been around 20 m across.

12 Quinn, *Roanoke voyages*, 215. It is not in Harriot's hand.

During these preliminary expeditions, the *Tyger* had been refloated, and the fleet of nine ships sailed north along the coast on 21 July, locating an inlet that became known as Port Fernando or Hatarask, just north of the present-day Bodie Island lighthouse; this provided a partially protected anchorage. The ships seem to have remained here while pinnaces were used to enter the shallow sound. Discussions were had with a local chief, Grangamineo, with Manteo acting as an intermediary, to locate a site for the main fort and settlement on Roanoke Island. Meanwhile Captain Amadas took a pinnace to explore the Albermarle sound and may have taken Harriot and White with him.

At this point, it seems that the original plans for the colony broke down, as it was clear that there was no safe anchorage for the larger ships, whose crews may have chosen the open ocean rather than the inadequate Port Fernando. One pictorial map, illustrated by de Bry, entitled 'The arrival of the Englishmen in Virginia' (and not included in the White watercolours), includes five shipwrecks and two vessels at anchor, with a pinnace working its way up the Albemarle Sound, possibly representing Amadas's expedition. Roanoke Island is shown with an Indian village, with a thin cover of trees but no location for the planned fort or settlement. As the map does not show the English settlement or fort on Roanoke Island, it may have been drawn up before the transfer to the island; while there is no record of the five shipwrecks, that may be largely schematic. The caption supplied by Harriot in 1590 describes the horrors of entering the sound and the first meeting with Indians on Roanoke Island. It was hardly good propaganda for the success of the colony.

One of the smaller vessels was dispatched to England on 5 August. At the time, preparations were being made for the other vessels to leave also, with Grenville and the *Tyger* setting sail on 25 August, leaving the all-male colonists to fend for themselves under the leadership of Ralph Lane, with only small boats in which to move around. One reason for Grenville's sudden departure may have been the lure of capturing Spanish prizes on the route home – which he did.[13] It is likely that White returned home with Grenville in August 1585; his name is not included in the list of colonists under Ralph Lane – unlike Harriot, who is listed. The details of White's watercolours could all have been obtained in the first few months of Grenville's expedition around the Pamlico Sound to the Indian villages of Pomeioc and Secotan, and it is notable that he does not illustrate any places visited after August 1585. His folios do, however, include maps, among them areas that were not explored by September 1585. These maps were surveyed by Harriot, and later redrawn by White in London in

13 Santa Maria Vincenza. This was sailed back to Grenville's home port of Bideford. A group of cannons, discovered when the waterfront was being repaired in the early twentieth century and now relocated around a bandstand, may have come from this ship.

1586–87. Harriot's activties in Virginia remain undocumented for the next nine months, despite his being named as a member of the colony. Gradually, the colony's fortunes collapsed, under pressure from conflict with the Indians and general starvation, and the survivors were eventually rescued by Francis Drake's returning fleet from the West Indies on 18 June 1586.

Sadly, Harriot's detailed Chronicle for these nine months has been lost, but the surviving maps give a good indication of the places he visited and surveyed. These suggest that Harriot's mapping accompanied Lane's expeditions up the Albemarle, Roanoke, and Chowen rivers, and maybe as far north as the Chesapeake Bay. At the same time, with his knowledge of Algonquian, he developed a first-hand ethnographic knowledge of the region, culture, and natural history.

John White was able to return to the New World, unlike Harriot, in 1587. He was appointed the governor of the settlers' colony established in July on Roanoke Island. But he spent only five months there, and left for England in November to secure new supplies. When he returned in 1590, he found the 'Cittie of Ralegh' abandoned. The only clue to the fate of the colonists was an arborglyph carved on a tree: the letters CRO, with another clue on the palisade, CROATAN, evidently the island of Croatoan, now Hatteras island, which was the ancestral village of Manteo, the key ally and supporter of English colonization.[14] White was unable to reach Croatoan, and so the myth of the Lost Colony was born.

Thomas Harriot as ethnographer

Harriot and the survivors of the colony arrived back in Portsmouth on 28 July 1586. Quinn has argued that Harriot then spent his time collecting together an archive of accounts of the expeditions up until 1586, which he passed to Richard Hakluyt for publication as well as to help him compile his now lost Chronicle.[15] It is probable that John White was working up his drawings at the same time and that he and Harriot collaborated in drawing up the maps from the original surveys. The plan might have been to provide a favourable view of the American 'paradise', despite the failures of the Grenville and Lane projects, and so to generate public support and investment for a settlers' colony, to be led by John White, newly appointed as Governor of the newly created Company of the Cittie of Raleigh in January 1587.

---

14 Quinn, *Roanoke voyages*, 613–14.
15 No trace of Harriot's Chronicle survives, although reference is made to it in *A briefe and true report*, 32–33: 'I have ready in a discourse by itself, in maner of a Chronicle according to the course of times, and when time shall bee thought convenient shall be also published'. See also Quinn, 'Thomas Harriot and the New World', 42.

Harriot compiled his *Briefe and true report of the new found land of Virginia* in 1586–87, completing the manuscript shortly after the new Company had been formed.[16] Curiously, it did not appear as a printed pamphlet for another year, until February 1588. The text was then reproduced by Richard Hakluyt in the 1589 edition of the *Principal navigations*. The printed version that is now well known was the first volume for a series entitled *America*, issued by the Frankfurt publisher Theodor de Bry. This included Harriot's text, with engravings and maps based on John White's originals, including some that are now lost. It is likely that Hakluyt brought White, Harriot, and de Bry together in London in late 1588 to develop the project with editions projected in French and Latin as well. Harriot added extended captions to the engravings; these further amplify his original text and most closely reflect his ethnographic observations. Thus the published version of 1590 comprised three elements: Harriot's original 1587 report, which may have been intended as an optimistic prospectus for the new Company but which was published after its departure; engravings based on White's original drawings; and extended captions to these drawings, written by Harriot in 1588–89, after White had returned from America but before his failed attempt to locate the colonists and his own family in 1590.

Harriot's report begins with descriptions of the commodities. For two items, he records the local name: wapeih (a medicinal soil) and winauk (sassafras). In his description for iron, he states that he found this on the waterside, 'the ground to be rocky, which by the trial of the mineral men was found to hold iron richly' at a distance of 'about four score miles and the other six score miles' from the fort. There are recorded haematite ore beds outcropping close to the banks of the Roanoke River, now known as the Gaston Beds, in Halifax County.[17] Nineteenth-century descriptions of the deposit fit closely to Harriot's; they are located 120 miles from Roanoke Island, with beds up to 8 ft thick. The mention of the 'mineral men' refers to Joachin Gans and his fellow mineralologists.

Harriot also describes copper, an item of particular interest to the Elizabethans, whose development of bronze cannons made an American

16 Quinn, 'Thomas Harriot and the New World', 46–47 explains the complexity of production, proposing that the text was first written immediately after his return in 1586 as a prospectus for the 1587 expedition, but that it was held back for publication for some reason until February 1588 and then reprinted by Hakluyt a year later; this version is in Quinn, *Roanoke voyages*, 317–89. The de Bry edition was then produced in 1590. It is available in facsimile as Thomas Harriot, *A briefe and true report of the new found land of Virginia*, with a new introduction by Paul Hulton (New York: Dover, 1972). The pagination in my references is that of the original 1590 text, reproduced in the Dover text.

17 H. B. C. Nitze, *The iron ores of North Carolina* (Raleigh, 1893), 43.

source especially valuable.[18] He had observed inhabitants wearing diverse small plates of copper in two towns 150 miles 'into the main'. As he noted, the Indians traded these from further away, where 'they say are mountains and rivers that yield white grains of metal, which is to be deemed silver'.[19] He also records Indians wearing silver. The English were able to obtain some of this copper, tested it and found it contained silver. Native Americans were unable to smelt copper or silver but mined it in its native form. While much of the copper trade was from the Great Lakes region, there were also deposits around the upper reaches of the Roanoke River in Mecklenburg and Halifax county, Virginia, around 180 miles from Roanoke.[20]

The second part covers 'such commodities as Virginia is known to yield for victual and sustenance of mans life'. Many of the plants were unknown to Harriot, so they are unusually listed by their Algonquian names, with comparisons to the European equivalents such as gourds, peas, and beans, as well as the staple of maize. He includes details of the ground preparation with digging sticks, planting, and harvesting. Tobacco, known as vppówoc, is described as a herb. Dried and powdered, the fumes or smoke was then inhaled 'through pipes made of clay into their stomach and head, from whence it purges superfluous fleam and other gross humours'. He comments that the Indians sucked it after their manner, 'as also since our return', suggesting that some tobacco was brought back to England.

The narrative continues with descriptions of roots and fruits, mostly using the local names, such as coscúshaw, probably the arrow-arum, which is correctly observed as partly toxic, and sagatémener, which are chinquapin nuts which he notes were first dried and then, when used, rehydrated and turned into a flour. His description of animals includes deer, rabbits, grey squirrels, bears, and two animals to which he gives the local names of saqenúckot (marsh rabbit?) and maquówoc (mink?). He records 28 different names of animals, but only identifies twelve of them through observation; this is clear evidence that he must have transcribed as many names as he

---

18 Elizabethan concern to control the supply and manufacture of copper and brass led to the creation of the Company of Mineral and Battery Works (1565) and Company of Mines Royal (1568). Limited supplies of copper were mined in Cumberland and Cornwall and smelted near Keswick and Neath in South Wales, but the Company of Mines Royal encountered technical difficulties and a lack of capital investment. New supplies of copper would have been very desirable; see M. B. Donald, *Elizabethan copper. The history of the Company of Mines Royal 1568–1605* (Ulverston: Red Earth, 1994). Ralegh was made Warden of the Stannaries in 1585 and would have had an interest in minerals.

19 Harriot, *Briefe and true report,* 10.

20 https://www.dmme.virginia.gov/webmaps/DGMR/. Lane describes his expedition up the Roanoke River in search of this mine at Chawnis Temoatan, which he did not reach because of the removal of the colony in 1586. He adds that 'For this river of Moratico (i.e., Roanoke River) promiseth great things, and by the opinion of Master Harriots, the head of it, by the description of the country either riseth from the Bay of Mexico or else from very near unto the same, that open into the South Sea'; Quinn, *Roanoke voyages,* 273–74.

could in the course of direct conversation. As for birds, he records the local names of 86 species, though he had only observed eight types of waterfowl, and seventeen land fowl. He also notes fish, but in less detail; while unable to record their species, he comments that twelve examples had been drawn, complete with their local names.[21] He describes night fishing (where fire was carried in the canoes, and the light used to attract the fish), barbed spears, and fish traps, set out in the shallow waters of the sound. The account of food ends with descriptions of shellfish and turtles.

The third section covers, first, building materials, detailing the different types of timber available. Again, Harriot records many of the names in the local dialect, but does not list them, as 'seeing for timber and other necessary uses, I have named sufficient'. He concludes with the observation that 120 miles from the fort a 'gentleman of our company' had located by the river, a 'great vein of ragge stone', presumably a hard limestone or sandstone.

There then follows the ethnographic section, 'of the nature and manners of the people', covering six pages of detailed observation. He begins with descriptions of their clothes, bows, and clubs before moving onto their houses and villages, which contained ten or twelve houses, occasionally twenty, the greatest number being thirty. The houses, up to 12 or 16 yards long, were made from small poles, bent over in the manner 'used in many of the arbories in our gardens of England'.[22] The village centred around the wiróns, or local chief, who might control one village, or up to eight; but the largest number he recorded under a single chief was 18, and could muster a fighting force of 800. He observes that dialects changed between villages, 'the farther they are distant the greater is the difference', and notices their craft: 'considering the want of such means as we have, they seem very ingenious'.

The real ethnographic detail, however, is found in Harriot's extended captions to the engravings based on John White's drawings. Here de Bry uses the originals as a basis, but often adds details taken from Harriot's account. For example, the town of Secotan is engraved with additional plots of tobacco, pumpkins, and sunflowers, described in the accompanying captions as well as the inhabitants hunting deer with bows and arrows. The illustration of fishing is likewise embellished with additional species of fish. There has been debate whether some of the engravings that do not have equivalent watercolours, are from lost paintings or were made up later in London from the existing material and Harriot's descriptions – for example, the description and illustration of how they made dug-out canoes.[23]

---

21 In the Sloane volume of watercolours in the British Library (BL Add. MS 5263) there are watercolours of North Carolina birds and fish, with Algonquian names. These are assumed to have been drawn by White. See Sloan, *New world*, 230–32.

22 *Briefe and true report*, 24.

23 Sloan, *New world*, 108.

One set of drawings shows religious practices, what is likely to be the Green Corn ceremony in the depiction of the village of Secotan, which is shown in detail in both paintings and engravings as a separate drawing, with the description 'solemn feast whereunto the neighbours of the towns adjoining repare from all parts'. The ceremony involves dancing around wooden posts carved with heads, with three virgins embracing in the centre; at the conclusion they retire to a place sounding rattles, while set around a fire, as a fire ceremony. Another shows the mortuary house of the chiefs, where ten bodies are laid out; according to Harriot, the guts and bowels had first been removed, then the flesh was removed under the skin, before the bones were re-covered with additional leather coverings. A fire was kept nearby, with a wooden idol of their deity kiwasa guarding the corpses. The engravings also include an image and description of *kiwasa*, wearing a necklace of copper beads and sat in a curious tent, which is not found in White's drawings.

The most celebrated of John White's drawings are of the inhabitants. Here de Bry and Harriot often add a rear view of the same figure and set them within the landscape of the Outer Banks, showing villages, fishing and hunting scenes, and wildlife. Two of the most striking are the illustration of the 'cheiff Lorde of Roanoac' and the 'cheiff Ladye of Pomeiooc'. The male chief has been identified as Wingina, whom the English dealt with (and killed in the summer of 1586). The figure shows both a necklace of copper and pearl beads, also a rectangular piece of copper suspended around the neck. The female image shows a woman carrying a large gourd and includes a child with a doll and a rattle; Harriot comments that the children are 'greatly delighted with puppets and babes which were brought out of England'. Three further plates describe cooking fish on a rack over an open fire, boiling a stew of 'fruit, flesh and fruit' in a large ceramic pot, which in the paintings has a conical point as a base,[24] and eating hulled maize from a wooden platter. This engraving adds extra objects not shown in the painting, but which Harriot describes, including a smoking pipe,[25] a gourd, a tobacco pouch, a fish, corn cobs, and a scallop shell.

Thomas Harriot as cartographer

Harriot's first involvement with Ralegh's circle was to provide navigational instruction to his sea-captains, in the course of which he constructed a substantial manuscript, now lost, the *Arcticon*. While he must have helped guide Grenville's fleet to the Outer Banks, his role on the expedition seems also to have been to compile accurate maps. Three of these have survived, two as watercolour drawings in the hand of John White, and a third, lost

---

24 The shapes of the pots are exactly as are found in archaeological deposits of this date.
25 This may be the oldest representation of a tobacco pipe in England, and is very similar to those found on the Outer Banks at this date.

original, included as an engraving by de Bry. Between them, these maps provide a detailed guide to the areas visited by Harriot, and an insight into his cartographic skills.

Two maps are entitled Virginia Pars.[26] Virginia Pars one (VP1) shows the coast from Florida to the Chesapeake, with drawings of sea creatures, and the progress of the *Tyger*, as well as a scale of latitude. This map is compiled from the two sources in London: material from La Moyne's drawings when he was part of Ribault and Laudonnière's expedition to Florida (1564–65), including some of the place names; the section of the map south of Cape Lookout is somewhat schematic.[27] Northwards the map's surveying is based on a measured triangulation survey rather than astronomical observation, Harriot having taken his latitude observation in the area of Cape Lookout at the start of his survey, as is suggested in the table below.

Location	Latitude	Harriot's Latitude	Difference (n. miles)
Cape Lookout	34 37	34 30	7
Cape Hatteras	35 13	35 50	37
Chicamacomico	35 33	37 00	87
Roanoke Island, N End	35 56	38 05	129
Curritack Lighthouse	36 22	39 30	172
Cape Henry	36 55	40 40	225

It has been observed how accurate the maps are if overlain over a modern map. Now we have satellite imagery and mapping software not available to earlier writers, we can interrogate the maps more closely. If VP1 is overlain for best fit on the satellite imagery, the latitude bar, for each degree (60 nautical miles), measures only 23 miles. The orientation, as shown by the compass rose, is 355 degrees, or 5 degrees west of due north.[28]

Virginia Pars 2 is a more detailed watercoloured map. No latitude bar is given, but there is a scale and compass rose. By georeferencing it onto the modern map, the orientation is 10 degrees east of north; the scale shown in leagues; 10 leagues as 29.51 nautical miles, very close to the 30 miles. The orientation error is around 10 degrees east of north, close to the predicted

26 Shown in reasonable detail in Sloan, *New world*, 94–95 and 106–7.
27 Le Moyne fled to England after the St Bartholomew's Day massacre in 1572, and ended up in Walter Ralegh's circle, as an accomplished botanical illustrator. See Paul Hulton, 'An album of plant drawings by Jacques le Moyne de Morgues', *British Museum quarterly*, 26, nos. 1/2 (1962), 37–39.
28 It is interesting that there is such variation in the orientation of these maps. Harriot wrote about how to correct for magnetic variation using the rising and setting of the sun (BL Add. MS 6789, ff. 534–37). Jon V. Pepper, 'Harriot's earlier work on mathematical navigation: theory and practice', in Shirley (ed.), *Thomas Harriot*, 54–63.

error of magnetic declination, modelled for Cape Lookout in 1590, of 12 degrees.[29] The discrepancy between the two maps might suggest that VP1 was compiled without Harriot's direct supervision, and that the information was 'fitted' into what was believed to be the shape of the North American eastern seaboard, without due regard to scale or orientation.

Virginia Pars 3 is only found in the engraved version of de Bry and repeats many of the details of VP1 and VP2, but includes more details of the Chesapeake and the Chowan and Roanoke rivers, suggesting the there was a third, now lost original. It is scaled and orientated, but the scale of 25 leagues should read 45 miles on the ground but is actually only 36 nautical miles in length[30] with the orientation of 5 degrees east of north. While schematic in places, these rivers are shown with details sufficiently convincing to suggest that Harriot managed to survey a considerable distance inland following the rivers and to enter the Chesapeake for some distance by sea, possibly even reaching the James River. This is not surprising, as in his written account he describes villages up to 120 miles inland, reached by travelling up both rivers. VP2 and VP3 are much less clear on the southern area of the Pamlico sound, with the Core Sound, Neus River, and Pamlico River only shown schematically; these must have been visited in July 1585, with Grenville and White, before the formal survey work had commenced further north around Roanoke Island. All three maps were probably compiled back in England, using the survey data that Harriot recorded in the field, with White converting the manuscript maps into watercolour paintings.

Harriot gives an insight into the survey instruments he was carrying, when he records the reaction of the Indians to his equipment, 'that they thought were more the work of gods than of men'. He lists 'as mathematical instruments sea compasses, the virtue of the loadstone in drawing iron, a perspective glass whereby was showed many strange sights,[31] burning glasses, wildfire works,[32] guns, books, writing and reading, spring clocks that seem to go of themselves, and many other things that we had'.[33] Surveying instruments that might have been available to Harriot have been discussed in the context of a list compiled in 1582 for a proposed survey of the coastline of New England, by Thomas Bavin.[34] These include a cross staff, a sailing compass, a table (presumably a plane-table), and two pairs of brazen compasses. This list does not include the theodolite, or any form of linear measuring device.

Harriot's survey methods would have included compass bearings, dead-reckoning, and triangulation. The accuracy of VP2, which was probably the

29 https://maps.ngdc.noaa.gov/viewers/historical_declination/
30 It is possible that the engraver converted this to land leagues, so the distance was 40 miles.
31 Most likely a telescope.
32 Possibly fireworks.
33 *Brief and true report*, 27.
34 E. G. R. Taylor, 'Instructions to a colonial surveyor', *Mariner's mirror*, 37 (1951), 48–62.

'master map', could only have been achieved through triangulation, given the limitation of astronomical observations and the measurement of longitude. Triangulation was well known at this time and was explained in Thomas Digges's *Pantometria*, published in 1571.[35] Harriot's knowledge of trigonometry enabled him to make direct calculations, rather than rely on analogue drawing.[36]

The Outer Banks are low-lying islands, which rapidly disappear below the horizon. The islands themselves were densely forested, making any ground-based survey very difficult, so that Harriot must have undertaken his survey mostly by boat. His first task would have been to lay out an accurate measured baseline along the shoreline, where each end was visible from the sound. Then from his boat, he would have taken angular measurements between the two ends of the baseline, and a third or fourth position, before moving the boat to a new station, taking further angular measurements and working his way around the sound. There is no information as to the device used – whether it was an early type of theodolite, or a device known as a nautical compass (Bavin's 'brazen compass') consisting of two brass arms each with a sight, with an measured arch between them, so the angle between two points could be directly measured.[37] The inclusion of fireworks among his equipment might suggest that they were used to locate ground stations over longer distances, possibly in conjunction with a theodolite, set up on dry land. The Atlantic coastline might have been measured in a similar way, but more likely by dead-reckoning, using compass direction and speed over the water, although having ascertained the location and shape of the island and inlets, it would have been straightforward to estimate the ocean shoreline. Distance and direction estimation would have been used as the surveyors moved up the rivers, as the basic shape is correct, but their location on the ground is more variable; for example, the fork in the Roanoke river, clearly observed by the surveyors, is located 45 miles further south than its true location.

One feature of the maps is the careful location of the Indian villages, of which 31 are noted, most of them by name. VP3 shows them as small rings of stakes, while VP2 shows the villages with red dots. but also includes an annotation in a pink wash to suggest that those areas had populations that were friendly to the English.

35 Thomas Digges, *A geometrical practice named Pantometria* (London, 1571).

36 For example, Harriot papers (Petworth House), 403–453, p. 11. From the first baseline measured triangle, further lengths can be calculated using the sine rule, from measured angles. See also Harriot papers (Petworth), vol. VIb: The doctrine of nautical triangles.

37 Two English examples dating from the late sixteenth century are in the Dudley collection, Museo Galileo (https://catalogue.museogalileo.it/object/NauticalCompass_n01.html and https://catalogue.museogalileo.it/object/NauticalCompass.htmlThere is also an early theodolite, dated 1590, in the Dudley Collection. But this would have been much more difficult to use on board a small vessel such as a pinnace, and was probably not necessary as the coastline is so low-lying (https://catalogue.museogalileo.it/object/Theodolite_n04.html).

Thomas Harriot and archaeology

Interest in locating the English activities in North Carolina goes back to the beginning of historical archaeology in America, with a hope of finding remains of their forts and settlements and resolving the mystery of the 'Lost Colony'. Alongside this and often unrelated, have been studies of the contemporary Native American cultures, termed Late Woodland, which provide a parallel source of information to Harriot's ethnographic observations.

While earlier discoveries had been made on Roanoke Island, the first 'archaeological' investigations by Talcott Williams, in 1895, took place at an earthwork that had become known as Fort Raleigh.[38] Williams was drawn to the low earthworks and ditches in woodland in the northeastern part of the island. His methods were somewhat crude, but he did map the fort and excavated thirteen trenches inside and in the nearby woods. Within the fort, he recorded a layer of humus, with an ashy layer below containing some iron but nothing diagnostic of English settlement. After Williams's investigations, the site and its archaeology were compromised, first by its use for a film set that involved digging a trench and rampart through the fort in 1921, and then by the construction of a replica log fort in 1938. By 1948, the log fort had become rotten, and the decision was taken to remove it, enabling two years of excavations by J. C. 'Pinky' Harrington in 1949–50.[39] He stripped the site to recover an accurate plan, while removing most of the remaining interior deposits. In the process, a tiny number of English artefacts were found which could be argued to be sixteenth century in date.[40] Not believing that he had found either the main fort or the settlement, Harrington then excavated a further 48 trenches across the park[41] and

---

38 The designation of the earthworks as Fort Raleigh probably dates from 1819, when President James Monroe was shown 'the remains of the Fort, the traces of which are still distinctly visible which it is said to have been erected by the first colony of Sir Walter Raleigh'; quoted from the *Edenton gazette* in G.C. Grassl, *The first English settlement in America* (Bloomington, 2006), p. 10.

39 J. C. Harrington, *Search for the Cittie of Raleigh. Archaeological excavations at the Fort Raleigh National Historic Site, North Carolina* (National Parks Service, 1962).

40 An iron sickle found in the ditch fill, a gauge auger, a number of wrought iron spikes or nails, three Nuremberg counters (whose precise find spot is unknown), a small brass balance weight, two pieces of copper casting waste, two musket balls, a glass bead, 22 sherds of Spanish olive jar, probably from the same vessel, one piece of redware, two majolica sherds, and a piece of clay roof tile; Harrington, *Cittie of Raleigh*, 17–22. In 1965, Harrington returned to investigate a 'brick path' that proved to be a timber building, and four curved furnace bricks within a pit.

41 Very few artefacts were discovered, but four fragments of triangular metalworking crucibles were found 200 feet northwest of the fort. A brass finial was also recorded in the roots of a fallen tree, 50 feet west of the fort, which Harrington suggested was the top of an andiron. While the identification is correct, it is more likely to date to the eighteenth century, where similar examples are in the Williamsburg Foundation collections; Harrington, *Cittie of Raleigh*, 36–37.

adjacent Elizabethan Gardens.[42] There, the only feature of interest was a wooden stick-lined pit filled with charcoal, 100 feet west of the fort, which he suggested was a charcoal burning pit of unknown date. After he completed the excavations, the fort was completely reconstructed in 1950, with banks and ditches dug out to form an earthwork that approximated the excavated plan, for the public to view. Fort Raleigh, as it is visited today, dates from 1950.

Matters remained thus for some years, with no sign of the settlement or indeed Lane's fort.[43] Re-examination of Harringtons's stratigraphic record in 1989 led Ivor Noël Hume to argue that some of the English material found by Harrington was actually sealed by the fort's ramparts, suggesting that the fort must date from *after* the Elizabethan occupation. As early as 1948, doubts were being expressed that the earthwork was actually Fort Pain, known to have been built in the same area during the French and Indian War (1754–63) and shown on maps since 1768.[44] The Park Service still maintains a sixteenth-century origin for the Fort. But, given its small size with an internal space of around 557 m^2,[45] it cannot be the main fort constructed by Lane, although it bears a superficial similarity in plan and size to the small entrenchment for storing salt built in Puerto Rico, drawn by John White.[46]

In 1989–92, Noël Hume and William Kelso were given permission to excavate to the west of the fort as well as within the fort itself.[47] The impetus was to investigate a square posthole structure that resembled the tower

---

42 A privately owned site, open to the public, recreating Tudor-style gardens. Fort Raleigh and the adjacent woodland is a National Park.

43 Quinn was convinced that the settlement and fort were located at Fort Raleigh, positing a cove, long lost to the sea, and a bizarre hypothetical plan, on the basis of no archaeological evidence! Quinn, *Set fair*, pp. 409–10.

44 In NPS files held by NPS historian Charles W. Porter, quoted in Grassl, *First English settlement*, 58. Captain Paine commanded the North Carolina Provincials during the war and was operating around the Outer Banks. The suggestion still remains controversial, and when I discussed this with Noël Hume, shortly before his death, he was less certain of his earlier conclusion. Lawler, *Secret token*, 148–50 captures some of present controversy around Fort Raleigh, and our meeting with Noël Hume.

45 Roughly 75 ft by 80 ft (internally 23 × 24 m), according to Harrington's plan. The size of the Puerto Rico fort can be judged from the size of the boat being drawn in. It was around 20 m internally.

46 The discovery of the Jamestown and Martin's Hundred forts on the James River, dating to the early seventeenth century, show that they were constructed as palisades rather than as bank and ditches. This concurs with the description of the fort sought out by John White for the survivors of his colony in 1590.

47 The excavations of Noël Hulme and Kelso are reported in a manuscript 'First and lost: in search of America's first English settlement', but were never published; file in the NPS library at Fort Raleigh. A popular account was published in the *Colonial Williamsburg journal* and reprinted in Ivor Noël Hume, 'Roanoke Island, America's first science center', in *In search of this and that* (Williamsburg, 1996), 96–109; see also Nicholas M. Lucketti, 'Copper Carrieth Ye Price Of All, or how Thomas Harriot may have saved Jamestown', in Eric Klingelhofer (ed.), *A glorious empire. Archaeology and the Tudor-Stuart Atlantic world* (Oxford, 2013), 1–11.

found at Martin's Hundred on the James River, which was recorded by Harrington in 1965, in the hope that there was an earlier phase of timber fort. What was actually found was a scatter of English artefacts in and around the fort, including the rim of a glass alembic jar, a piece of antimony, a fragment of a ceramic Normandy flask, sixteenth-century Delftware bottles, and small metalworking crucible fragments. Taking the finds with those found earlier by Harrington of crucible fragments and a find in 1848 of 'two glass globes containing quicksilver', which are now lost, the conclusion was that these artefacts were remains of the mineralogical laboratory. This was thought to have been set up by Joachim Gans in 1585, no doubt with Harriot's assistance, to examine the mineral ores of iron, copper, and silver which he says he collected in the interior.[48] The square posthole structure observed by Harrington is likely to have formed part of the complex, possibly their workshop. The finds and the charcoal pit were over a substantial area, suggesting that these activities were scattered across maybe 100 m². Gans and Harriot would have found a location well away from the settlement to undertake their experiments, which involved noxious gases but also required charcoal-burning to fuel the furnaces.

Subsequently only scattered finds have been made at Fort Raleigh, despite extensive survey and excavation in the vicinity of the fort. One reason is that much of the site was covered in dune sand after a hurricane in 1769. Heavy erosion of the coast has led to speculation that the main site of both Lane's fort and the later colony have been washed away – a notion supported by the discovery of a timber barrel-well in the sound just east of the Lost Colony theatre. Fragments of Spanish olive jars were found in the beach dunes, where a thin cultural horizon is visible. Excavations in 2006 in Prince Town woods revealed no further English material.[49] But in 2016 a small project in the same area located a single sherd of a Delft pharmaceutical jar, probably associated with industrial activity.

In 2008–9 a large excavation, undertaken by the First Colony Foundation in the 'Harriot nature trail' around 122 m to the west of Fort Raleigh, produced further evidence of industrial activity: scattered artefacts of European origin, a Normandy Martincamp flask, and a crucible sherd,

---

48 Gans is mentioned in the list as Dougham Gannes, and later as Master Yougham; see Quinn, *Roanoke voyages*, 196, 274. Gans was a Jew from Prague, but was in England in 1581, at the Keswick works of the Mines Royal Company. On his return from the New World, he is recorded at Bristol (maybe on his way to Neath furnaces in south Wales), where he was charged with blasphemy in 1589; Quinn, *Roanoke voyages,* 196. Grassl, *First settlement*, 221–43 provides a lengthy biography for Gans. It is quite simple to tell whether copper is of local origin, due to the absence of arsenic, of which English copper had considerable quantities. The piece of spongey copper from Roanoke has not been tested.
49 Nicholas Lucketti, 'Archaeological survey of Prince House Woods Fort Raleigh National Historic Site Roanoke island, North Carolina', Unpublished report for National Parks Service (2007).

suggesting the material from the metallurgical activity may have been widespread. Two pits were also found, one with a few Venetian glass beads and two brass aiglets, a second containing an articulated necklace made of twelve copper squares of decreasing size. While undated, analysis of this necklace showed it was made from English copper, and thus likely a trade item, quite possibly given to the Roanoke Indians by the colony, who brought it with them as a diplomatic present.[50]

The absence of hard archaeological evidence for either settlement and forts at Fort Raleigh has led to speculation that Lane's fort and White's colony may be located elsewhere. In the absence of any cartographic evidence, the main locational evidence comes from John White's attempt to find his abandoned colonists in 1590. He describes how he took two boats up from Port Fernando/Hatorask, making for where the planters were left. They overshot the place,[51] by a quarter of a mile, but then saw a light at the north end of the island and carried on rowing. They sang English songs, and there was no answer, so they landed, found the remains of the fire, made through the woods to the western side of the island,[52] and returned by the waterside, round the north point of the island, until they reached the place where White had left the colonists in 1587. There was a sand bank (sand dune?) on which the famous tree was located with the CRO initials carved into the trunck. White then continued on to locate the remains of the houses where he had left the colonists, but which were now taken down. He noted that the place was fortified by a 'high palisade of high trees' that must have been erected after his departure, on which the full word CROATOAN had been carved. He then carried on along the shore, towards 'the point of the creek'. There they found some of their sailors reporting that they had located where the chests had been hidden, at the end of an old trench made two years past by Captain Amadas. This

---

50 Eric Klingelhofer and Nicholas Luccketti, 'Elizabethan activities at Roanoke', in Peter Pope and Shannon Lewis-Simpson (eds.), *Exploring Atlantic transitions* (Woodbridge, 2013), 181–89; Carter C. Hudgins, 'Copper, chemistry and colonisation: the roles of non-ferrous metals at Jamestown (c. 1607–10) and Roanoke (c. 1585–90)', ibid., 202–14; Luccketti, 'Copper carrieth', 7. Analysis of the copper confirmed its European origin. The 1584 expedition had already noted the value of copper to the native Americans, and Harriot would probably have discussed this with Manteo and Wanchese; Quinn, *Roanoke voyages*, 102–3. One of John White's paintings also shows a touching scene of the wife of an Indian chief and her daughter, carrying an English doll and pointing to her necklace with double strands of gold and copper, likely to have also been a gift; Sloan, *New world*, 122. Harriot commented, in the caption to the de Bry engraving, that 'they are greatly delighted with puppets and babes were brought out of England'; *Briefe and true report*, plate VII. Both pits may have been dug by the Roanoke Indians with a deliberately placed deposit of goods they had obtained from the English; any organic deposit are unlikely to have survived.
51 By which he means landing place, likely to be the area known as Shallowbag Bay, but in the sixteenth century a much smaller creek.
52 It is rather unclear why they crossed the island at this point.

find included five chests, three of which belonged to White; they had presumably been buried for safety.[53]

This passage is difficult to interpret precisely, but the implication is that White's settlement was closer to Baum Point than Fort Raleigh,[54] somewhere at the end of the run of dunes along the north coast and the present Shallowbag Bay. The coastline has changed radically since 1590 due to a sea-level rise of about 60 cm since the 1590 s[55] and massive coastal erosion that has removed around 500 m of shore.[56] Fortunately, there is an accurate survey of 1716 which gives a better indication of the shape of the island, only 125 years later, with no Shallowbag Bay but Gibson Creek with marshland to the southeast, now flooded as the open water of the bay.[57] When White mentions the 'point of the creek', he must surely be referring to the head of this creek, where the sailors unearthed the chests nearby. The 'old trench' where they were found may have been part of Lane's original fort, as Amadas was part of the Lane colony, and where the sailors had gone to check at a place that they already knew. It is likely that White's settlement and Lane's fort were close to each other. As both colonies relied on boats for their transport; such a position lies near an anchorage, where the pinnaces could be protected. The fort at Puerto Rico was a D shape, around 80 m in size, with an unfortified section alongside a similar creek.

This area northwest of Gibson Creek/Shallowbag Bay is known as Mother Vineyard, after five vines growing here, estimated to be 400 years old. It is a scuppermong vine, a variety of muscadine grape (*Vitis rotundifolia*) indigenous to North America. The vines are planted in neat rows, and while there is no proof, it is tempting to suggest that Harriot might have planted them close to the 1585 settlement. He was aware of the local vines

53 The contents included 'his books torn from their covers, the frames of some of my pictures and maps rotten and spoiled with rain'; Quinn, *Roanoke voyages*, 615.

54 Quinn was convinced that Fort Raleigh was the main site, and so tried to make the geography fit by believing the point of the creek was Otis Cove, immediately east of Fort Raleigh. This spit was apparently created in the early twentieth century by placing obstacles along the beach; see Grassl, *First English settlement*, 97.

55 R. E. Kopp, B. P. Horton, A. C. Kemp, *et al.,* 'Past and future sea-level rise along the coast of North Carolina, USA', *Climatic change*, 132 (2015), 693–707. https://doi.org/10.1007/s10584-015-1451-x.

56 Recent studies of the north shore of Roanoke Island show an erosion rate of between 3 and 8 feet a year. At an average of 4 feet a year, this would lose 524 m since the 1590s, and Fort Raleigh would be 700 m from the shore. See Stanley R. Riggs and Dorothea V. Ames, *Drowning the North Carolina coast. Sea-level rise and estuarine dynamics* (Raleigh, 2003), 102–7.

57 A land map compiled by E. Moseley, 1729, but a 'true copy' of a map surveyed by W. Maule in 1716; Harrington, *Cittie of Raleigh*, 7, fig 6. On White's map VP2, a very small inlet is shown where Shallowbag Bay is today; also shown on the perspective map of the island in de Bry. By the time of Collet's 1768 map, the creek is shown, but a little larger. The present shape of the bay is recorded on nineteenth-century maps, and this may be the result of the 1769 hurricane.

and their economic potential, and he noted two types of grape, 'one small and sour ... the other far greater & of himself lushious and sweet'.[58]

The archaeology of Hatteras Island

John White was convinced that the remnants of the 1587 colony had left for Croatoan Island, as a result of the message carved on the tree and the palisade; he concluded also that they had left voluntarily, as neither was 'without a cross or sign of distress'.[59] The use of these symbols was described as a 'secret token', agreed between the settlers and White before he left for England in 1587. It seems the plan was for the colonists to 'remove from Roanoke 50 miles into the main'[60] after White's departure, hence the need to mark a tree as to their destination. Whether some of the colonists actually went inland has been the subject of intense historical and archaeological conjecture.[61] But the arborglyphs clearly show that some, if not all, went to Croatoan Island, abandoning their palisaded settlement on Roanoke Island.

Croatoan Island is now named Hatteras Island, as the inlets through the Outer Banks have been reconfigured in the last 400 years. Their reason for choosing this location was probably two-fold. The island was home to Manteo, one of the loyal Algonquians, who was a key guide and interpreter to the English, and a place where they could expect protection and hospitality. The island's Cape Hatteras projects into the Atlantic, so they could spy any passing ship, with the hope of being rescued. White had passed the island on his way to Roanoke in 1590; he anchored offshore, and even sent a small boat to take soundings around the inlet and the shoals lying off Cape Hatteras.[62]

---

58 *Briefe and true report*, 9. There is also a tradition that a tree stump located close to Shallowbag Bay was the remains of the famous tree with the CRO inscription. Harrington, *Cittie of Raleigh*, 54.

59 Quinn, *Roanoke voyages*, 614.

60 Quinn, *Roanoke voyages*, 613. One issue has been the interpretation of the 'main', which the OED gives as short for mainland, as used in the sixteenth century. White elsewhere writes 'between the main (as we supposed) and the island', showing that he meant the mainland rather than islands; Quinn, *Roanoke voyages*, 609. The most logical meaning must be 50 miles away from Roanoke on the continent, which could include the Chesapeake or up the Roanoke or Chowen rivers. The point at which these rivers flow into the Albemarle is almost exactly 50 miles distant from Roanoke.

61 Horn, *A kingdom strange*. More recently the First Colony Foundation have been working in Bertie County, at the head of the Albemarle Sound, where the Roanoke and Chowen rivers meet. The interest in this area was kindled by the discovery of a patch on the Virginia Pars map, hiding an icon for a fort. As the map was drawn after the 1585–86 expedition, it could well have shown a base for exploring these rivers, which for some reason had to remain secret. Recent announcements by the FCF claim sixteenth-century artefacts linked to the fate of the Lost Colonists.

62 Quinn, *Roanoke voyages*, end map, attempted to plot the actual soundings. It remains a mystery why the colony's survivors, only shortly arrived, would not have observed this activity.

Hatteras island is one of the few stable areas of the Outer Banks with a land area wide enough to support coastal forest and fertile soils. The inlet into Pamlico sound, where the fresh and seawater mix, provides a rich marine ecosystem, while the shallow sound is rich in fish and shellfish. The archaeological record indicates that the island was continuously occupied from the Middle Woodland (200 BCE–500 CE) until the present day.[63] Our excavations at the Jeanette Creek site showed an Algonquian occupation until c. 1740, alongside a pioneer English settlement of the same date, whose descendants still live on the island today.

Archaeological research on the island dates to the 1950s, when Smithsonian archaeologist William Haag undertook a comprehensive survey of the Outer Banks and Pamlico sounds, locating twelve sites on the island and noting in particular the Cape Creek site (31DR1), at the western end of the island. In the 1980s and until 1998, David Phelps, from East Carolina University, worked on the island and conducted annual seasons of excavations at Cape Creek from 1992 to 1998. Among Phelps's discoveries in the largely seventeenth-century Cape Creek site was a signet ring, believed to be gold, but actually of brass, and part of a snaphaunce mechanism that dates to the late sixteenth century. These were seen as possible evidence of the English on the island. Another stray find, made in the centre of the island in the 1950s, was a Nuremburg counter of the same type as found at Fort Raleigh.[64] It was unclear whether these items could have come from any of the English expeditions or, given their context, whether they were trade items acquired from the English, maybe from seventeenth-century Jamestown.

We were invited to follow up on these discoveries, working with the Croatoan Archaeological Society. Since 2009, we have worked on a number of sites on the island, including the Jeanette Creek site,[65] the Hatteras School site, and between 2012 and 2018 at Cape Creek itself, which was threatened with redevelopment for housing. A substantial part of the site had already been redeveloped in 2006, with landscaping and earthmoving, but we were fortunate to be able to work in areas that had not yet been disturbed, except by the previous excavations of David Phelps.

---

63 Stone tools from the island include Clovis, Cumberland, and Savanna points, suggesting a low-level palaeo-Indian and archaic presence. The only absent period is Early Woodland (1000–200 BCE).

64 According to Haag, coins and a counter were found close to 35 13′ 50″/75 37′38″ by a man called Tandy and handed over to a 'government man'. The counter is in the NPS store; illustrated by Harrington, *Cittie of Raleigh*, 20. The fate of the coins is unknown. See W. G. Haag, *The archeology of Coastal North Carolina* (Baton Rouge, 1958).

65 Louisa Pittman, 'The myth in the memory. Towards a new archaeology of Hatteras Island', Ph.D. thesis (University of Bristol, 2014).

The Cape Creek site covers around 0.5 ha or 100 × 50 m, bisected by a shallow creek. A dune ridge forms the northern edge of occupation, which falls off towards the sound, suggesting that the occupation, represented by thick, dark midden deposits, was on the protected south side of the dune ridge. The middens were then covered in a further 60 cm of wind-blown sand – possibly from the 1769 hurricane, that may also have closed the inlet – before being stabilized by coastal forest cover of mostly Live Oak.

The site has two cultural horizons, A and B; one dated to c. 1550–1600, and a second c. 1640–1700. West of the small creek, only the older A horizon was found to comprise dark brown occupational deposit, while to the east of the creek, the A horizon comprised patches of dense shell midden, with the later dark occupation soil above. Both occupation deposits contained multiple postholes for timber buildings.

Most of Phelps's excavations, as well as our own, uncovered a rich mid seventeenth-century deposit, and it was here that the ring and snaphaunce were found, along with a coin of Charles II. We found three coin-weights, all dating to 1648 but pierced so they could be hung on a string, and probably traded by an English merchant, as well as gun flints and lead shot manu-facturing waste and part of a gun barrel. There were dress accessories such as a hook and eye closure, and a decorative fitting for leather, as well as dress pins, suggesting that European clothes were being worn. Two pieces of glass, one of sufficient quality to have come from an instrument such as a compass, had been reused as artefacts. Quantities of copper-cutting waste show that copper sheet was being worked, and Phelps had already found a copper figurine and several furnaces to work or anneal the copper plate. Chemical analysis of the copper indicated it was imported from Europe. Two items of importance stood out. The first was an iron sword hilt with traces of gilding, of a type common in the late sixteenth century – and thus a second example, along with the snaphaunce, of high-status weaponry that could be associated with the Elizabethan settlement. The second item was a fragment of writing slate, on which were drawn multiple lines, clearly used for sketching. Taken together, these items could be explained as the survival of heirlooms in this mid-seventeenth-century settlement. However, the bulk of European material was traded to the island, presumably in exchange for local commodities such as skins and fur.

The earlier horizon was largely devoid of English material: the shell middens east of the Creek completely so, but to the west, four items were found. The first was the neck of a stoneware vessel, of a type current in the late sixteenth or early seventeenth centuries; the second a section of bar iron of the type used in trade; the third was a plano-convex bun of pure copper, such as would be recovered from a small experimental smelt. Perhaps the most interesting find was another Nuremburg counter, the second from Hatteras Island, and comparable to the three from Roanoke Island, and the fourth known previously from Hatteras. This group have been attributed to either Hans Schultz II or III, who were operating between 1584 and 1612.[66]

The excavations have been especially important in providing an insight into Harriot's ethnographic world and the changes that took place through European contact. Analysis of the faunal assemblages provide an even mix of bird, deer, fish, turtle, and small land mammals, while in the seventeenth century, deer and land mammals come to dominate the fauna, presumably as a result of the introduction of firearms and a trade in animal skins. Bird bones also increase in number, as firearms replace nets. Before contact, there was heavy reliance on shellfish – mostly small clams and scallops, but also whelks, periwinkles, and large ocean clams. Some of their middens often contain a rich mix of shellfish and turtle and deer bones, but others just contain shells. There is little evidence of seasonality, showing that they lived on the island all year around. The plan of their houses was difficult to establish in many of the excavations as there were multiple postholes, from regular rebuilding. However, one house was exceptionally clear, at least 5 m in length and 3 m wide, with rounded ends, as drawn and described by Harriot and White.

Harriot as ethnographer

Thomas Harriot's formative experience was during his period in the New World in 1585–86, as a 25-year-old, full of youthful energy to discover the world. While the Grenville-Lane expedition has had a poor reputation, owing to its cruel treatment of the local population and its ultimate failure and evacuation, this was not due to Harriot but to the military arrogance of its leaders. In contrast, in the work of Harriot and White, one can see a humanism[67] and tolerance, and evidence of scientific and ethnographic curiosity

---

66 There were three generations of Hans Schultzes producing tokens: HS I (1553–84), HS II (1584?–1603), and HS III (1608–12). One example from the ditch of Fort Raleigh and the Cape Creek example is HS III (rose and orb type), while the two others found near to the fort and the stray find from Hatteras island are HS II; Michael Mitchiner, *Jetons medalets & tokens. Vol. 1. The medieval period and Nuremberg* (London, 1988), 404–11. Beverley Straube, '"A sure token of their being there": artefacts from England's colonial ventures at Roanoke and Jamestown', in *Exploring Atlantic transitions*, 190–202 contends that the tokens were all part of a later trade from Jamestown, but HS III tokens are rare from there. As these tokens are undated, Mitchiner's attribution is made on typology, and it is curious that the only time Roanoke and Hatteras were connected was during the English activities, and it is quite possible that the rose and orb type was actually produced by Hans Schultes II, whose dates would fit well. Straube notes that HS II and HS III jetons were found in deposits dating to 1590 s from Bankside, London, and it may be that the typology needs revision; Julian Bowsher and Pat Miller, *The Rose and the Globe Playhouses of Shakespeare's Bankside, Southwark. Excavations 1988–91* (London 2009), 216.

67 Harriot and White's 'humanism', from where their careful ethnographic observations were inspired, are discussed in Andrew Fitzmaurice, *Humanism and America. An intellectual history of English colonisation 1500–1625* (Cambridge 2003), 50–57.

about the world that they were working in. What we now see as ethnography was probably viewed as expediency to survive and exploit this strange world.

Often overlooked, the first engraving of his *Briefe and true report* is not based on John White's paintings of native culture. It is an image of the Garden of Eden, with a very European Adam and Eve taking the apple from the tree, beasts in the foreground, and agricultural scenes behind. There was nothing explicitly American about it; de Bry was probably reusing the engraving. But the message is clear. What follows was a kind of Garden of Eden, and we should conclude with Thomas Harriot's own sympathetic words from the *Briefe and true report* that were so unusual for the time:

> Although (friendly reader) man by his disobedience, were deprived of those good gifts wher with he was indued in his creation, yet he was not berefte of wit to provyde for hym selfe, nor discretion to devise things necessarie for his use, except suche as appartayne to his soules healthe, as may be gathered by this savage nations, of whom this present work intreateth. For although they have noe true knowledge of God nor of his holye word and are destituted of all lerninge, yet they passe us in many things, as in sober feedinge and dexteritye of witte, in makinge without any instrument of mettall thinges so neate and so fine, as a man would scarsclye beleve the same, unless the Englishmen had made proofe thereof by their travails into the contrye.[68]

---

68 Harriot, *Briefe and true report*, 'To the gentle Reader', preface to engravings.

5 Harriot, Hakluyt, and the *Briefe and true report ... of Virginia*

Daniel Carey

Thomas Harriot's *Briefe and true report of the new found land of Virginia* stands as one of the most remarkable records of the English encounter with the New World. It has been the subject of extensive and longstanding critical discussion as a founding text in the effort to establish a colonial presence in North America, including Harriot's depiction of indigenous Algonquian peoples, his assessment of natural resources, and his analysis of the commercial potential of the territory's commodities. While the volume appeared as a separate, stand-alone publication, in 1588, printed by R. Robinson, as well as in the famous edition produced by Theodor de Bry in 1590 with its engravings based on the work of the illustrator and colonial agent John White, it had a third incarnation which provides the focus of my discussion. The great English editor and exponent of travel, commerce, and expansion, Richard Hakluyt, reprinted the work in 1589 as part of *The principall navigations, voiages, and discoveries of the English nation*, and then once more in the final volume of his expanded edition of the collection in 1600 (*The third and last volume of the voyages, navigations, traffiques, and discoveries of the English Nation*).

The question I want to explore is what we can learn by treating Harriot's work as an embedded text, that is, by reading it alongside other narratives of the attempt to settle in Virginia printed by Hakluyt in his compilation and chronicle of the voyages that took place under the patent of Walter Ralegh. Here it forms part of a series of contributions exhibiting a considerable diversity of attitudes and a narrative arc that complicates our reception of what Harriot has to say. My argument is that the distinctiveness of Harriot's viewpoint emerges more clearly in this context, enabling us to appreciate his particular preoccupations and sensibility. I will follow the historical sequence as Hakluyt presents it, except that I intend to preserve discussion of Harriot until the end in order to illuminate the strategies of his *Briefe and true report*. This progression will take us through a sequence running from the narrative of the first Virginia voyage by Arthur Barlowe in 1584 to the experience in 1585-86 of the governor and military commander Ralph Lane, followed by the descriptions provided by the agent and artist John White of journeys made in 1587 and 1590. Hakluyt places Harriot's *Briefe and true*

DOI: 10.4324/9781003096580-6

report in chronological succession after writings by Lane and before those of White – appropriately given that he was part of Lane's expedition – but its meaning is inevitably inflected by the materials that surround it.[1] By taking this alternative approach, we can gain a new perspective on the structure, argument and significance of Harriot's work.

Barlowe, Lane, and White on Virginia

In his presentation of the Virginia voyages,[2] Hakluyt began by printing Ralegh's patent, which established the scope and legality of the venture, before turning to the description of the first voyage of reconnaissance captained by Philip Amadas, a member of Ralegh's household, which took place in 1584. The text was composed by Arthur Barlowe, a participant in the expedition who was also styled a captain.[3] Barlowe's vivid prose sets a tone of optimism, and highlights, above all, opportunity and cooperation as keynotes of the experience of the undertaking. This befits the text's purpose as a promotional document submitted to Ralegh (who in turn, presumably, provided it to Hakluyt who enjoyed his patronage). Barlowe addresses Ralegh expressly as his target audience, together with Elizabeth I, and what he refers to as the 'Common wealth' (92). The evidence of the opening indicates that the document derived from the ship's log, which provided a narrative structure for the outward voyage, in particular, but Barlowe designates it as a 'discourse' and eschews the tedium of what he calls the 'diurnall of our course, sailing thither, and returning' (92). It represents something of a hybrid text to the extent that it holds at times to a chronological order but departs from this frame for more extended descriptions that fill out a wider picture of resources, ethnographic observation, and important narrative interlude. Barlowe clearly wants to capture a full sense of the occasion, including the encounter with indigenous people, description of what the land yields, and its potential.

1 The timeline is as follows: (1) the reconnaissance voyage in April 1584 under Philip Amadas and Arthur Barlowe, landing in the Roanoke territory in July and departing September 1584, described in Barlowe's narrative; (2) the voyage of Sir Richard Grenville in April 1585, establishing the so-called first colony under the governorship of Ralph Lane in June. The group includes Thomas Harriot and the artist John White. Grenville returns to England. The settlers depart in June 1586 with Francis Drake; (3) in April 1587 an expedition led by John White sails for Roanoke with a new group of settlers forming the so-called second colony; he returns to England in August; (4) John White returns to the territory in August 1590 in search of the settlers left behind, arriving back in England in October after failing to make contact.
2 I have quoted throughout from the texts as edited by David Beers Quinn, *The Roanoke voyages 1584–1590*, 2 vols. (London: The Hakluyt Society, 1955), continuously paginated. References to quotations are included in the text.
3 Little is known about Barlowe. Quinn describes him as one of the 'subordinates in the company that Ralegh commanded in the Irish wars in 1580-81' with previous experience of a voyage to the eastern Mediterranean. David Beers Quinn, *Set fair for Roanoke. Voyages and colonies, 1584–1606* (Chapel Hill: University of North Carolina Press, 1985), 22.

Barlowe's heightening of the sensual experience of the territory begins even before arriving on terra firma. On 2 July 1584, the mariners found shoal water that 'smelt so sweetly, and was so strong a smell, as if we had bene in the midst of some delicate garden, abounding with all kind of odiferous flowers' (93–94), portending a near approach to land that they would soon claim, with formal ceremonies, for the English crown. The territory itself, on their first observation, abounded with grapes, which offered themselves in plenty: 'the very beating, and surge of the Sea overflowed [with] them' (95). Nowhere else was like it, even those parts of Europe – as he had seen for himself – renowned for their store of grapes.[4] The difference, he acknowledged, was 'incredible to be written' (95). Thus from the start the encounter is presented as exhilarating, with a poetic cadence that signals fecundity, sweetness, and abundance – a kind of seduction, in other words.

He frames the experience initially as an encounter with the land itself, partly by keeping to a chronological record, that is, before making contact with the native inhabitants.[5] The valleys are 'replenished' with cedars, and the woods full of deer, rabbit, and fowl 'in incredible aboundance' (96).[6] Abundance is certainly a watchword, picked up in Hakluyt's marginal annotations.[7] Barlowe introduces comparisons with European and extra-European places to communicate a favourable image of the territory, notably its supply of sweet smelling trees. It is not until the third day that native peoples are encountered. The first meeting occurs with a group of 40 or 50 men, led by Granganimeo, brother of the local king (as Barlowe designates him) Wingina.[8] Barlowe comfortably interprets the signs of this

---

4 The emphasis on grapes draws attention to the fertility and bountifulness of the land, while appealing to the appetite. Images of grapes appear frequently in Renaissance art in association with feasting and in conjunction with religious iconography. See E. de Jongh, 'Grape symbolism in paintings of the 16th and 17th centuries', *Simiolus. Netherlands quarterly for the history of art*, 7, no. 4 (1974), 166-91.

5 'We remained by the side of this Island two whole daies, before we saw any people of the Countrey; the third daye we espied one small boate rowing towards us ...'; Quinn, *Roanoke voyages*, 97–98.

6 For his repetition of the notion of replenishment with respect to cedars see also Quinn, *Roanoke voyages*, 115.

7 Richard Hakluyt, *The principal navigations, voyages, traffiques, and discoveries of the English nation*, 3 vols. (London, 1598–1600), vol. 3, 247. Quinn, *Roanoke voyages*, 98, n.3. On this theme, see Rachel Winchcombe, *Encountering early America* (Manchester: Manchester University Press, 2021), 180-81.

8 Quinn, *Roanoke voyages*, 98. On Wingina and Granganimeo, see Michael Leroy Oberg, *The head in Edward Nugent's hand. Roanoke's forgotten Indians* (Philadelphia: University of Pennsylvania Press, 2008). Wingina is described by Barlowe as the king of Wingandacoa (Quinn, *Roanoke voyages*, 113), an ill-defined territory which either included the town and surroundings of Secotan or an alliance with its peoples. Quinn refers to him as the chief of the Roanoke tribe; *Roanoke voyages*, 99, n. 2. Lee Miller, *Roanoke. Solving the mystery of the Lost Colony* (New York: Penguin, 2002), 265-69, argues that Wingina led the Secotans. Oberg emphasises the 'fluidity of Algonquian community politics' (169).

ceremonial, ascribing joy and welcome to Granganimeo's gestures and even describing his thoughts with the same assurance: he 'neuer mistrusted any harme to be offered from us' (99). This episode is the prelude to the opening of a profitable trading relationship, clarifying the kinds of goods available in the country and the items that the English possessed that proved in demand, notably a bright tin dish.[9]

Barlowe's report is not merely a narrative but also an ethnography to the extent that he explains the social values of the people, their respect for nobility and rank, the king's authority, the costume of the natives, the deportment of women, their practices in making boats, method of agriculture and harvesting, diet, worship, weapons, tools, and war-making. The account is not systematic; part of its appeal is that information is intermingled with narrative as Barlowe attempts to convey the totality of the experience and its meaning for readers. He builds towards an image of an accepting people, whose simplicity of life engages the observer and echoes with classical and biblical associations:

> We were entertained with all loue, and kindnes, and with as much bountie, after their manner, as they could possibly deuise. Wee found the people most gentle louing, and faithful, void of all guile, and treason, and such as liued after the manner of the golden age. The earth bringeth foorth all things in aboundance, as in the first creation, without toile or labour. (108)

The immediate connection is clearly with Ovid and the invocation of the Golden Age near the opening of the *Metamorphoses* (1.89–112) where the land affords its produce without human effort.[10] Barlowe's simultaneous reference to the 'first creation' introduces a further Biblical resonance with a prelapsarian condition. Interestingly, Hakluyt omitted that sentence when he printed the text in 1600.[11] Was he trying to steer the reader away from a delusional expectation that no effort was required to survive in the New

---

9 Quinn, *Roanoke voyages*, 100–101.
10 In Arthur Golding's 1565 translation the passage reads:

> Then sprang vp first the golden age, which of it selfe maintainde,
> The truth and right of euery thing vnforst and vnconstrainde.
> There was no feare of punishment, there was no threatning lawe
> In brazen tables nayled vp, to cronch [*sic*: crouch] or créepe to Iudge with cap in hand,
> ... .
> The fertile earth as yet was frée, vntoucht of spade or plough,
> And yet it yéelded of it selfe of euery things [*sic*] inough.

> Ovid, *The. xv. bookes of P. Ouidius Naso, entytuled Metamorphosis*, trans. Arthur Golding (London, 1567), B2r–v. Golding published the first four books in 1565.

11 Compare *The principall navgations* (1589), 731, with *The principal navigations*, vol. 3, 249, which elides the sentence: 'The earth bringeth foorth all things in aboundance, as in the first creation, without toile or labour.'

World? The absence of treason, consistent with Ovid's depiction of a time before laws and systems of authority, may speak in this context to the lack of threat to English command.[12]

As Barlowe's narrative progresses, he tells a story that casts a shadow over this idyllic scene.[13] He relates that some two years earlier the king of a rival group (Piemacum, *weroance* of Pomeiooc) called a party together that included his former foe, the lord of Secotan (subordinate to Wingina), ostensibly to confirm peace agreed between them, and then suddenly swept in among the guests and killed the men, sparing the women and children.[14] Thus Barlowe's narrative complicates or contradicts his own assessment – Ovid's emphasis on the peacefulness of the Golden Age is gone[15] – but he does not reconcile the difficulty. Virginia remains, to that extent, in suspense, making an invitation but colouring it with potential doubt (Figure 5.1).

After printing Barlowe's account, Hakluyt turns to the voyage made by Sir Richard Grenville in 1585, who set out for Virginia in April of that year and left behind a group of colonists under the governorship of Ralph Lane.[16] Lane's prior experience included various forms of military service in Ireland, such as building fortifications in the province of war-torn Munster, where he came to the attention of Ralegh.[17] Hakluyt prints two writings by Lane, the first being his letter to Hakluyt's cousin, Richard Hakluyt the elder, of the Middle Temple, dated 3 September 1585.[18] This short letter offers a straightforward celebration of Virginia, declaring it to be 'the

---

12 Barlowe tactfully omits reference to Ovid's inclusion among the virtues of the Golden Age that people kept to their own lands (*mortales praeter sua litora norant*) and had not yet chopped down the pine to build ships and take to the sea (*liquidas undas*). Golding's translation reads: 'The loftie Pynetrée was not hewen from mountaines where it stood, / In séeking straunge and forren landes, to roue vpon the flood. / Men knew none other countries yet, than where themselues did kéepe' (B2v).

13 On this point, see also Kathleen Donegan, *Seasons of misery. Catastrophe and colonial settlement in early America* (Philadelphia: University of Pennsylvania Press, 2013), 29–30.

14 Quinn, *Roanoke voyages*, 113-14.

15 In Golding's version: 'No horne nor trumpet was in vse, no sword nor helmet worne, / The worlde was suche, that souldiers helpe might easly be forborne' (B2v).

16 Lane's contribution remains neglected relative to that of Harriot and John White. For discussion, see Donegan, *Seasons of misery*, ch. 1; Michael G. Moran, *Inventing Virginia. Sir Walter Raleigh and the rhetoric of colonization, 1585–1590* (Bern: Peter Lang, 2007).

17 See Rory Rapple, 'Brazen as Falstaff, devious as Iago: Sir Ralph Lane's approach to officeholding in both Ireland and Virginia' forthcoming in *Journal of British studies*. Grenville was also experienced in Ireland. See David Loades, 'Grenville, Sir Richard (1542–1591)', *Oxford dictionary of national biography*.

18 On the possible misdating of this letter (it could be 8 September 1585), see Quinn's note, *Roanoke voyages*, 210, n. 2. In fact, in 1589 Hakluyt had to print the item out of sequence since it must have arrived to him late in the day; it appears in *The principall navigations* (1589), 793, but he reordered it in 1600, where it preceded Lane's longer description of the colony's experience from August 1585 to June 1586.

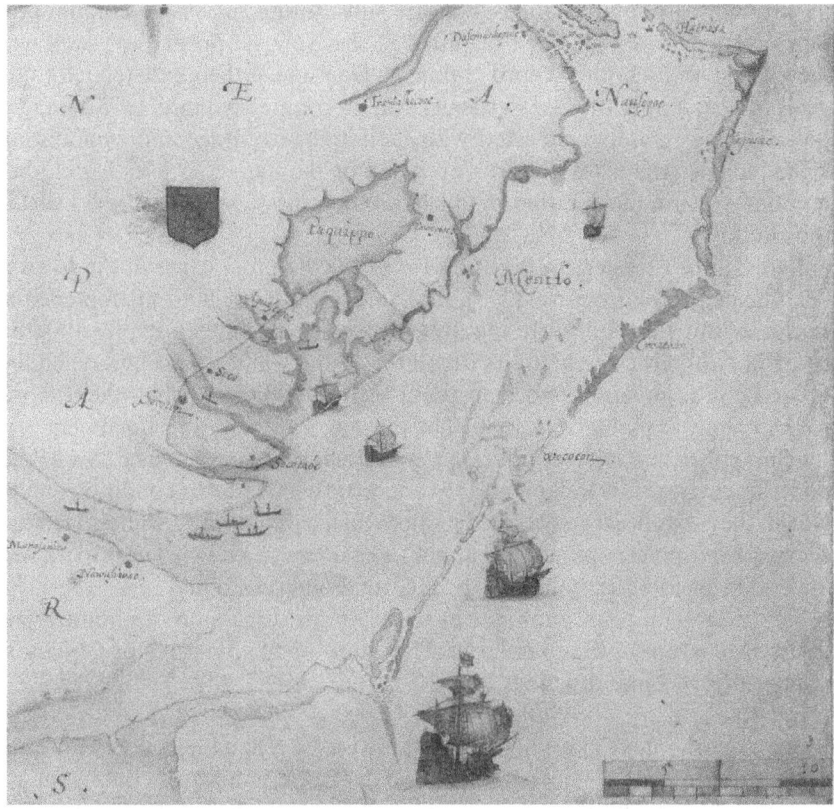

Figure 5.1 Detail of map by John White entitled 'La Virginea Pars', depicting the territory explored by Ralph Lane, including Secotan. Sir Richard Grenville's larger ships, the *Tyger* and the *Elizabeth*, appear at the bottom of the image. By permission of the British Museum.

goodliest soil under the cope of heauen' (207),[19] and highlighting a comparable roster of enticing goods to ones that Barlowe had recorded: gums, sweet trees, grapes, flax, silk grass, what he calls guinea wheat (identified as maize by Hakluyt), apothecary drugs, sugar cane, and terra sigillata or Samian earth. The climate itself was very healthful: 'no realme in Christendome were

19 Although the phrase 'cope of heaven' was in wider use, Lane may have drawn here on Sir George Peckham's treatise, *A true reporte, of the late discoueries, and possession, taken in the right of the Crowne of Englande, of the new-found landes: by that valiaunt and worthye gentleman, Sir Humphrey Gilbert Knight* (London, 1583), which affirmed of the territory under Gilbert's patent: 'The climate [is] mylde and temperate, neither too hotte nor too colde, so that vnder the cope of heauen there is not any where to be founde a more conuenient place to plant and inhabite in' (B3v).

comparable to it' (208). Exuberance aside, the framing of Virginia around its productive potential and scope for substituting expensive commodities purchased abroad with those acquired in the New World fits very well with the agenda set by both Richard Hakluyts. One side of their case related to the geo-political imperative of settlement as a counter-balance to Spain, but the other was closely connected with economic advantage and exploitation. Lane includes the ostensibly strong desire of the native peoples for clothes, especially coarse cloth rather than silk (209), a message close to the heart of the Hakluyts.[20]

Lane's much longer contribution (over 9,000 words) came in the form of what he called a 'discourse' (255), the same generic designation employed by Barlowe. But this is a more overtly constructed document, expressly structured in two parts, the first describing the country and the second defending his decision as governor to decamp with Sir Francis Drake before the resupply of the colony took place in 1586. The purpose served by this document is as much exculpatory as descriptive. It also reflects Lane's experience as a soldier and administrator in Ireland and his concern to narrate the conditions under which they explored the territory – as Ralegh required – while defending themselves from conspiracy and attack. The forms of knowledge at work here are conditioned by that military background and sensibility.

This point comes through, for example, in his discussion of an important figure, Menatonon, king (as Lane styles him) of the province of Choanoke (Chawanoac). Lane describes him as

> a man impotent in his lims, but otherwise for a Sauage, a very graue and wise man, and of very singular good discourse in matters concerning the state, not onely of his owne Countrey, and the disposition of his owne men, but also of his neighbours round about him as wel farre as neere, and of the commodities that eche Countrey yieldeth. (259)

After making this high commendation, Lane drops in without explanation that this valuable relationship was struck up while he held Menatonon captive for two days. During that time he learned more than he had gained from his other 'searches and saluages' (259). He acquired information, for example, about a territory ruled by a king possessing significant quantities of pearl. Menatonon offered to provide guides to lead Lane there.

We remain in doubt, nonetheless, about the circumstances of his taking Menatonon captive. We learn that when Lane released him he retained Menatonon's son Skiko as a pledge. Skiko in turn told Lane about

---

20 See G. D. Ramsay, 'Clothworkers, Merchants Adventurers and Richard Hakluyt', *English historical review*, 92 (1977), 504-21. Aspects of the argument need revision, but there is no question about Hakluyt's enthusiasm for creating markets and demand for English finished cloth.

potentially valuable stores of a 'marueilous and most strange Minerall' (268), similar to copper, held by a certain people to whom access could be gained by Menatonon. Lane wanted to obtain samples to conduct an assay. To find out more, his plan on the expedition was to take one or two of these people hostage. Once again, intelligence and discovery are here underpinned by military tactics.

What can we derive from this part of the discourse? For one, English prospects were conditioned by intelligence gained from native sources. The whole affair depended on understanding alliances, treacheries, territories and their leaders, all of which made accessing natural resources difficult and complex. This was not the yielding territory imagined by Barlowe and celebrated in initial Lane's letter. It was highly mediated and contested, with human relationships defining much of what was possible. In this respect, it must have reminded him of his Irish experience and also depended on it to some extent for his understanding of risk and how to manage affairs in a hostile environment (Figure 5.2).

The second part of Lane's discourse describes the conspiracy of Wingina[21] – the *weroance* or chief of Wingandacoa[22] – and Lane's decision to depart with Drake on the homeward voyage. Once again we see how much the experience depended on precarious friendships and assistance, undercut by stratagems, rumour, and shifting alliances. The English had one major ally, Ensenore, described by Lane as a 'sauage father' (275) to Wingina.[23] His place in the story confirms the crucial human and character-led dimension of the colonial experience. Everything changed when Ensenore died on 20 April 1586. Wingina took up with a range of enemies against the English. He engineered a conspiracy by allying with other weroances, intending to execute Lane and then dispatch the rest of the English, including Harriot.[24] As part of his plan, he would break up the weirs, forcing the English to disperse in order to find stores of shellfish. Once again, Lane came to acquire intelligence of all of this from a captive, on this occasion Menatonon's son, Skiko, who 'flatly [discovered] all to me' (285).[25]

21 John White's drawing labelled by him 'A chiefe Herowan' may depict Wingina. For a reproduction, see Kim Sloan, *A new world. England's first view of America* (Chapel Hill: University of North Carolina Press, 2007), 139. Wingina later took the name Pemisapan. Karen Ordahl Kupperman suggests that this was Wingina's war name, assumed when he decided to attempt the destruction of the first colony; *Roanoke. The abandoned colony* (Totowa, NJ: Rowman and Allanheld, 1984), 76. See also Michael Leroy Oberg, 'Between "Savage man" and "Most faithful Englishman": Manteo and the early Anglo-Indian exchange, 1584–1590', *Itinerario*, 24, no. 2 (2000), 156-57.

22 The Algonquian name for the region was Ossomocomuck. On these names and their possible meanings, see Quinn, *Roanoke voyages*, 117, 853-54.

23 Quinn asks in a note: 'A foster-father, stepfather, or, even, father-in-law?'; *Roanoke voyages*, 275, n. 7.

24 Quinn, *Roanoke voyages*, 282.

25 For an intriguing reading of this episode as an orchestrated plan by Menatonon, see Seth Mallios, *The deadly politics of giving. Exchange and violence at Ajacan, Roanoke, and Jamestown* (Tuscaloosa: University of Alabama Press, 2006), 71-72.

Figure 5.2 John White watercolour entitled 'A cheife Herowan', possibly depicting Wingina. By permission of the British Museum.

The end for Wingina came about through some complicated manoeuvrings by Lane. He paid a visit to Wingina who had seven or eight weroances or leading men with him. Lane gave the watchword 'Christ our victory' and set upon them. Wingina was shot but escaped. An Irishman, Edward Nugent, who served Lane, and the deputy provost set off in pursuit and returned with Wingina's head.

Curiously these dramatic events do not lead to the sequel: as David Quinn points out in his edition of the text in *The Roanoke voyages*, Lane does not explain what happened next, either in relation to Skiko or to the Indian forces that Wingina had planned to range against Lane and the English.[26] This is another instance of narrative lacunae and storytelling for specific purposes, which, once served, allow the narrative to continue in a separate direction. In any case, a week later Lane received word from Drake of an offer of supplies. The remainder of the account focuses on Drake's arrival (fresh from his assault on Cartagena) and liberality, terminating in the decision to join his fleet and not await the long-anticipated arrival of Grenville.

Harriot's *Briefe and true report* comes next in the sequence of writings compiled by Hakluyt (reprinted from the edition of 1588), but I will leave discussion of the work until the concluding section in order to draw attention to its distinctiveness and strategies of argument in comparison with the surrounding materials.

The remaining documents relate to what Hakluyt refers to as the second colony, under the governorship John White, who departed from England in April 1587.[27] White had earlier accompanied the first Grenville expedition in 1585 along with Harriot, during which he produced the drawings for which he remains famous (serving as the basis for the widely distributed de Bry engravings that accompanied the 1590 edition of Harriot's *Report*).[28] White's account of the 1587 expedition and colony established under his rule is divided into monthly headings, indicating the debt of the piece to the journal form. After the arrival of the party in Virginia, the story is bound up with ascertaining what became of the 15 men left behind by Grenville in 1586. No sign of this group was found other than the bones of one man 'which the Sauages had slaine long before' (524). The evidence for the claim that he died at the hands of natives remains unclear at this point. We only discover the source of this knowledge later when White reunites the interpreter Manteo with his family on Croatoan.[29] Manteo learned from them that a group of Indians, including the Secotans, joined forces and set upon the English group with intrigues and ambushes. A dramatic scene of skirmishing is reconstructed

26 Quinn, *Roanoke voyages*, 288, n. 1.

27 Hakluyt also includes a short prior document, possibly of his own composition (in any case not a first-hand account), that narrates two expeditions to resupply the colony in 1586, the second led by Sir Richard Grenville which arrived not long after Lane and the other settlers had departed with Drake. He deposited 15 (or possibly 18) ill-fated men whose task was to 'retaine possession of the Country'; Quinn, *Roanoke voyages*, 480. No trace of them was found by White in 1587. See Donegan, *Seasons of misery*, 57–60.

28 See Michiel van Groesen, *The representations of the overseas world in the De Bry Collection of Voyages (1590–1634)* (Leiden: Brill, 2012), 112-16.

29 On Manteo, see Oberg, 'Between "Savage man" and "Most faithful Englishman"'; and Alden T. Vaughan, 'Sir Walter Ralegh's Indian interpreters, 1584–1618', *William and Mary quarterly*, 59, no. 2 (2002), 341-76, esp. 341-52.

by White on the basis of native testimony, noting the advantage the assailants gained by surprise, setting houses alight, and exploiting the trees for defensive purposes as they engaged in battle.[30] A number of Englishmen escaped by boat, but their fate was uncertain and he presumed them dead. As we can see from this instance, much of White's discussion is governed by uncertainty, reminding us the colonial terrain was a space of contested or precarious knowledge.

White concludes his piece by explaining the circumstances in which he departed from the colony for England in order to obtain additional supplies. This part of the work parallels Lane's need to engage in elaborate self-justification. The decision was not straightforward for White. Although he alludes to the birth of his granddaughter (the first English child born in the colony, Virginia Dare), this fact seems not to have played a significant part in his reluctance to leave. He was troubled by the loss of reputation he would suffer by returning home so quickly, knowing that he would face accusations from slanderers charging that he had sailed to Virginia 'politickely' (533), that is, with no intention of remaining there. What seems to have exercised him just as much was the risk to his own goods left behind, either by spoilage or pilfering, potentially putting him in the position of being 'unfurnished' on his return to the colony. The assistants and others urged him to journey back all the same and made a bond in writing to preserve his goods or make up the loss. He even included a copy of the text as proof.[31]

The Virginia sequence concludes (in the third volume of *The principal navigations* (1600)) with White's narrative of the 1590 voyage, constructed once again on the basis of a journal proceeding by monthly intervals. The record is full of incidents passed over fairly quickly, like the capture of 'two young Saluages' (602) on Dominica who subsequently escaped on St Croix, or the story of a boy on board who ran away when they set down at Cape Tiburon, only to return ten days later almost starved. They found the bones of men who had evidently perished for want of food, either because they had 'stragled' from their company or had been deliberately landed there by 'some men of warre' (603). These episodes remind us of the commonness of situations of extremis, and they set the struggles of the colonists in a wider context of hardship, endurance, and death that featured so routinely in New World expeditions.

In August of 1590, the expedition at last arrived in Virginia. The construction of the journal chronologically ensures that the narrative maintains considerable tension as the search for the colonists left behind in 1587 begins. Much of this account is conducted on the basis of the interpretation of signs. Initially the appearance of smoke rising from the

---

30 Quinn, *Roanoke voyages*, 528-29.
31 Ibid., 534-35.

island of Roanoke gave White and his companions hope of discovering their people. On the way there, they saw smoke in another direction and sought it out, but this too proved fruitless: 'we found no man nor signe that any had been there lately' (611). Eventually they made their way to Roanoke Island and by trumpet call and singing of English tunes they hoped to make contact, again without success. When they reached the site, they noted footprints of 'Saluages feet' (613). On a tree at the top of a forested dune they observed 'curiously carued these faire Romane letters CRO' (613) which they took to signify the colonists' departure for Croatoan as an agreed secret token indicating their intentions. A cross had also been appointed as a further sign that they should make if they left in distress, but no such mark was found. In the area of fortifications they discovered another prominent tree with CROATOAN spelled out fully in capital letters, but once more without the agreed cross as sign of distress. In an old trench White came across five chests that had been hidden by the planters, among them three of his own. His belongings were strewn around:

> many of my things [were] spoyled and broken, and my bookes torne from the couers, the frames of some of my pictures and Mappes rotten and spoyled with rayne, and my armour almost eaten through with rust. (615)

White makes the inference that 'this could bee no other but the deede of the Sauages our enemies at Dasamongwepeuk' (615-16). Although he was grieved by the loss of his goods, he expressed joy due to the 'certaine token' of the colony's safe removal to Croatoan, where the resident 'Sauages' were friends (616). It is one of the curiosities of this famous episode, relating to what is known as the Lost Colony,[32] that White and his companions never made it to Croatoan to establish the existence of those they had left behind. They set sail for Croatoan but a series of misadventures with the anchor and cabling left them in a very bad condition, almost running aground. Foul weather, limited food, and a lack of fresh water convinced them to head away to supply their 'wants' in Hispaniola, Puerto Rico, or Trinidad where they would winter before coming back. In the end the winds were unfavourable and they made a course for England. The story that White offers is thus a tragic history of colonization, ending the narrative arc on a melancholy note that lacks satisfactory closure (Figure 5.3).[33]

---

32 For a review of research and conjecture on the fate of the colonists, see Audrey Horning, *Ireland in the Virginian sea. Colonialism in the British Atlantic* (Chapel Hill: University of North Carolina Press, 2013), 125-27.

33 For further discussion of White's experience, see Donegan, *Seasons of misery,* 60–68.

Figure 5.3 Detail of map 'Americae pars, nunc Virginia dicta', engraved by Theodor
de Bry for Thomas Harriot, *A briefe and true report of the new found land
of Virginia* (Frankfurt, 1590). Based on John White's map reproduced in
part as Figure 5.1. By permission of the John Carter Brown Library.

Harriot's Virginia

We can now turn to Thomas Harriot and his account of the *New found land
of Virginia*. Engagement with the texts that surround it in the telling of
Virginia – from Barlowe to Lane and White – calls into relief a number of
distinctive features of Harriot's approach.

One point of note is often overlooked. On the final page of the *Report*,
Harriot concludes his discussion by adverting to another literary project, a
chronicle of the affairs of Virginia that he had composed.[34] The disappearance
of this document represents an enormous loss, assuming he did in fact complete it as he suggests. But the reference to it signals something important

34 Quinn, *Roanoke voyages*, 387.

about the generic difference of the *Report*. It is expressly not a chronicle, not an historical summary of actions and defence of them.

What then is the *Report*? To pursue the matter negatively for a moment, it also clear that the work is *not* a journal and does not bear the marks of deriving from a such a source in any obvious way. Richard Hakluyt confronted this question when he included the work in his analytical table of contents. The table is not a simple listing of documents in sequence but rather configures items according to a number of categories. In this instance, Hakluyt places Harriot's account, interestingly, not in the 'Catalogue of the voyages and Navigations' relating to the colony but among a set of 'Letters Patents, discourses, observations, and advertisements belonging to the foresaid voyages made into Virginia', even though the running order restores it to a chronological sequence.[35] In this category, it keeps company with Lane's Discourse, but it obviously serves a different purpose. The main thing is that Harriot's text has little narrative dimension, nor does it take on the task of defending the political, nautical, or military decisions of Lane or Grenville. Hakluyt may have intended then to identify it under the loose category of discourses, observations, or advertisements. All three terms apply to the work in various respects.

The title tells us something – not just that it is brief and true but that it is a *Report*. A report suggests something presented as a source of information. Contemporary examples of titles from the second half of the sixteenth century indicate that 'true report' was something of a stock phrase, but the implication is that it represented the outcome of some kind of inquiry, imparting information as a result. The OED definitions based on contemporary usage are suggestive of an account derived from an investigation.

This is precisely Harriot's ambition – to provide information that would promote the possibilities of Virginia and to arrange the welter of experience acquired in the New World in such a way that coherent sense could be made of it for those most concerned in the voyage – his audience of participants in the enterprise, whether investors or planters. To do this, he explicitly shapes the text in a tripartite division: starting with the 'Merchantable commodities' (325); then the commodities available for what he calls 'victuall, and sustenance of mans life' (324); and lastly building materials. The discussion of the 'nature and manners of the people' (325) is treated as something of a coda.[36]

---

35 A.M and D.B. Quinn, 'Contents and sources of the three major works', in *The Hakluyt handbook*, ed. D.B. Quinn, 2 vols. (London: The Hakluyt Society, 1974), vol. 2, 440.

36 Harriot's 1588 title includes reference to the 'naturall inhabitants': *A briefe and true report of the new found land of Virginia: of the commodities there fond and to be rayed, as well merchantable, as others for victual, building and other necessarie vses for those that are and shalbe the planters there; and of the nature and manners of the natural inhabitants ...* . The de Bry edition abbreviates this to: *A briefe and true report of the new found land of Virginia of the commodities and of the nature and manners of the naturall inhabitants.* Hakluyt in 1589 and 1600 stands apart in omitting reference to the indigenous people: 'A briefe and true report of

Each section is then subdivided under a series of headings, whether a phrase or single word, allowing him to discuss particular commodities, make comparisons as required, and clarify the productive value of the entity in question. He starts the first section of merchantable commodities with silk grass, followed by silk worms; flax and hemp; alum; *wapeih* (an earth like terra sigillata);[37] pitch, tar, rosin, and turpentine; sassafras;[38] and so on. The second section on 'victual' follows the same format, although Harriot finds himself departing into longer discussion of native practices in agriculture.[39] He resumes course with his catchwords by introducing further subdivisions devoted to roots, fruits, beasts, fowl, and fish, often using native terms as his headings. His third section on building materials keeps to this structuring principle, concentrating on types of trees before discussing stone, brick, and lime.

The final section 'of the nature and manners of the people' is introduced with a note of reluctance or reticence: 'It resteth I speake a word or two of the naturall inhabitants' (368), a somewhat misleading remark given that the discussion proceeds for a substantial number of pages. Here some scope for narrative appears in the account. Yet for the most part Harriot adheres to his informational method. Although he does not use formal headings as such, it is clear that he runs through a series of topics in rendering the ethnography: he starts with dress, before moving on to their towns, houses, government, war-making, and religion or belief system.

Harriot's manner of imparting information would lend itself readily to presentation in the form of a Ramist table.[40] In other words he employs a structured relationship to observation and organization of knowledge. This says something about his educational background and conception of travel,

---

the new found land of Virginia: of the commodities there found and to be raised, as well merchantable as others' (748; repeated in 1600, vol. 3, 266).

37 Probably of interest for its supposed medicinal properties rather than as a clay for producing pottery. For a discussion of terra sigillata, see Arthur MacGregor, 'Medicinal terra sigillata: a historical, geological and typological review', *Geological Society, London, Special publications*, 375 (2012), 113-36.

38 On sassafras, see Russell M. Magnaghi, 'Sassafras and its role in early America, 1562–1662', *Terrae incognitae*, 29, no. 1 (1997), 10–21. Richard Hakluyt the elder's 1585 'Inducements to the liking of the voyage intended towards Virginia' states 'Sawed boords of Sassafras and Cedar, to be turned into small boxes for ladies and gentlewomen, would become a present trade'; *The original writings and correspondence of the two Richard Hakluyts*, ed. E.G.R. Taylor, 2 vols. (London: The Hakluyt Society, 1935), vol. 2, 335.

39 Quinn, *Roanoke voyages*, 241-42.

40 Among the extensive critical works on Petrus Ramus, see Mordechai Feingold (ed.), *The influence of Petrus Ramus. Studies in sixteenth and seventeenth century philosophy and sciences* (Basel: Schwabe, 2001). For his influence on travel method, see Daniel Carey, 'Inquiries, heads, and directions: orienting early modern travel', in *Travel narratives, the new science, and literary discourse, 1569–1750*, ed. Judy A. Hayden (Farnham: Ashgate, 2012), 25–51.

placing him in an emerging tradition across Europe of methodizing the practice and subjecting it to observational discipline.[41]

There are two sides to this equation. One is the recording of information in such a way as to give it order and intelligibility. The other side of it is that his journey was, in all likelihood, an instructed voyage. I mean that Harriot's observations were probably shaped by a set of instructions set out by the proponents of the expedition.[42] He may have developed them himself in advance based on reading and conversation with others, or received them as formal instructions. In any case, there are indications in various documents (some printed by Hakluyt) that instructions were routinely delivered to mariners. They vary in form and focus, with attention allocated to important issues such as surveying, taking soundings, charting coastlines, the keeping of journals, etc., and other instructions related to specific observations on commodities and requests for samples.[43] He would also have received different but crucial instruction in advance from Manteo and Wanchese, the two Algonquian men who came to England in 1584 on the return of the Amadas and Barlowe voyage, and from whom Harriot acquired his grounding in the language.[44]

---

41 For further discussion, see Daniel Carey, 'Hakluyt's instructions: *The principal navigations* and sixteenth-century travel advice', *Studies in travel writing*, 13, no. 2 (2009), 167-85. See also Daniela Hacke, 'Colonial sensescapes: Thomas Harriot and the production of knowledge', in *Empire of the senses. Sensory practices of colonialism in early America*, ed. Daniela Hacke and Paul Musselwhite (Leiden: Brill, 2018), 165-89 (esp. 170-73).

42 We have one surviving document of note with a close association with the expedition, 'For *master* Rauleys Viage', endorsed 'Notes geuen to M*aster* Candishe' (printed in Quinn, *Roanoke voyages*, 130-39), dating in all likelihood to not later than early January 1585. The document concentrates on fortifications and other military arrangements, as well as matters of government. But there is a recommendation to include in the group a physician to attend to the soldiers and to 'discouer the simpels of earbs plant*es* trees roothes, and stons', together with a 'good geographer to make discription of the lands discouerd, and w*ith* hym an exilent paynter, potticaris [apothecary]' (135). The author was evidently a professional soldier. Quinn suggests three possible candidates: Sir John Smythe, Sir Roger Williams, and Thomas Digges; *Roanoke voyages*, 20–22.

43 Carey, 'Hakluyt's instructions'. Harriot's own directions for the later Guiana voyage relate to nautical observation. See E. G. R. Taylor, 'Hariot's instructions for Ralegh's Voyage to Guiana, 1595', *The journal of navigation*, 5, no. 4 (1952), 45–50.

44 On Manteo and Wanchese, see Vaughan, 'Sir Walter Ralegh's Indian interpreters'. On Harriot's linguistic activities and accomplishments in Algonquian, see Vivian Salmon, 'Thomas Harriot (1560–1621) and the English origins of Algonkian linguistics', in Salmon, *Language and society in early modern England. Selected essays 1981–1994*, ed. Konrad Koerner (Amsterdam: John Benjamins Publishing Co., 1996), 143-72. Originally published in *Historiographia linguistica*, 19, no. 1 (1992), 25–56. Coll Thrush provocatively describes Manteo and Wanchese as co-authors of the *Briefe and true report*. See Thrush, *Indigenous London. Native travelers at the heart of empire* (New Haven: Yale University Press, 2016), 35. For further evidence of instruction that Harriot received from indigenous sources during his time in Virginia, see Ed White, 'Invisible tagkanysough', *PMLA* 120, no. 3 (2005), 751-67; and Kelly Wisecup, *Medical encounters. Knowledge and identity in early American literatures* (Amherst: University of Massachusetts Press, 2013), ch. 1.

In terms of formal efforts to instruct travel a remarkable instance appears in an anonymous manuscript, dating from 1582 or 1583. This text provided guidelines for reconnaissance in North America, associated with the plans for exploration by Humphrey Gilbert (whose patent Ralegh inherited). The opening page is missing. Had it survived we might know more about its provenance, but the remaining pages contain fascinating details on what to observe, combining the sort of directions particular to mariners with other points relevant to commercial potential based on a survey and assessment of the territory. The document refers to shellfish and where they are to be found; beasts; birds; cochineal; trees, fruits, and gums; metals and 'sub-myneralls'; herbs and apothecary drugs; pitch, tar, and rosin. The visitors are also to 'note the manner of their planting & manuring of the earthe'. A further direction called for attention to 'the dyversitie of their languages and in what places their speache beginnethe to alter' (with the man assigned this task asked to bring an English dictionary with him to record local equivalents). Separate instructions are provided for the surveyor and artist Thomas Bavin concerning astronomical observations, the variation of the compass, and things for him to map and draw. The final section makes reference to André Thevet's *Cosmographie universelle* (1575) and to Giovanni da Verrazano, summarizing commodities of note, ranging from trees to fruits, including 'grapes of sondrye sortes', foul, various animals, minerals, and precious stones.[45] David Quinn suggests that some of the instructions might have come from Gilbert himself, Sir George Peckham, Martin Frobisher, or either of the two Richard Hakluyts (elder and younger).[46]

Whether or not Hakluyt was involved in this particular piece, there are other cases where his interest in the value of providing instructions for voyages is clear. These too offer us an insight into the kinds of things that Harriot is likely to have been asked to observe or that he recognized as important in advance himself. In *Divers voyages* (1582), Hakluyt printed pertinent instructions devised by the elder Hakluyt for a New World voyage.[47] In this document, the elder Hakluyt starts by recommending attention to building materials necessary for settlement (lime; slates or, alternatively, clay for making tiles; stone for walls if bricks could not be fashioned; and timber), on the basis that without them 'no Citie may bee made nor people in civill

---

45 'Instructions for a voyage of reconnaissance to North America in 1582 or 1583', in *New American world. A documentary history of North America to 1612*, ed. David B. Quinn, 5 vols. (London: Macmillan Press, 1979), vol. 3, 239-45.

46 Ibid., 239. E. G. R. Taylor also suggested that William Borough might have been involved; 'Instructions to a colonial surveyor in 1582', *Mariner's mirror*, 37, no. 1 (1951), 48–62.

47 The elder Hakluyt's authorship of this document was not revealed until Hakluyt reprinted it in *The principall navigations* (1589). In *Divers voyages*, the source was only identified as a 'gentleman'. E. G. R. Taylor dates the document to 1578 and describes it as intended for Sir Humphrey Gilbert's voyage; *Original writings and correspondence*, vol. 1, 116.

sorte be kept together'.[48] Harriot himself included reference to these crucial resources, in part to address malicious remarks from those intent on spreading doubt about their availability in the territory.[49]

The elder Hakluyt goes on to list a series of potentially valuable commodities that might be secured, ranging from grapes for cultivation to olives, hides, and dyestuffs, adding that

> if great woods bee founde, if they be of Cypres, chests may bee made, if they bee of some kinde of trees, pitche and tarre may be made, if they bee of some other then they may yeelde Rosin, Turpentine, &c. and al for trade and trafficke ...[50]

No shortage of possibility existed but it required a determination to investigate, explore and exploit.[51] Harriot in a similar vein dwelt on materials for shipping, dyestuffs, furs and skins, oils, and, again like the elder Hakluyt, on the potential for growing sugar cane, quinces, oranges, and lemons.[52] While Harriot stated that he was unable to provide 'certain affirmation' of cypress in the region, he did point to the existence of cedar for making chests, boxes, bedsteads, lutes, virginals, 'and many things els' (364). The emphasis in the elder Hakluyt diverges from Harriot's insofar as he draws attention to identifying markets for English goods among indigenous peoples, a theme absent from Harriot's *Report*. But they share a preoccupation with commodities, for sustenance or for trade, as defining the field of vision.

This document in *Divers voyages* is immediately followed by an instructive item from Richard Hakluyt himself, under the title 'The names of certaine commodities growing in part of America', drawn from his reading of authorities including the French cosmographer André Thevet and travellers such as Verrazano, Cartier, Ribault, and Best.[53] The piece draws attention to individual commodities ranged under the following headings: beasts, birds, fishes, worms, trees, fruits, gums (rosin, pitch, tar, turpentine, etc.), spices and

---

48 *Original writings and correspondence*, vol. 1, 117.
49 Quinn, *Roanoke voyages*, 363.
50 *Original writings and correspondence*, vol. 1, 119.
51 There are shared points in the elder Hakluyt's 'Inducements to the liking of the Voyage intended towards Virginia' of 1585, including the prospect of shipping materials, dyestuffs, the potential for mining (gold, silver, copper, and iron), orange trees and lemons, and sugar cane; *Original writings and correspondence*, vol. 2, 327-32, 335. Harriot's *Briefe and true report* described iron, copper, and a metal deemed to be silver; Quinn, *Roanoke voyages*, 331-33.
52 *Original writings and correspondence*, vol. 1, 119; Harriot in Quinn, *Roanoke voyages*, 330-32, 334-36.
53 This grouping of authorities perhaps points to the responsibility of Hakluyt for the final section of the 1582 or 1583 instructions printed in Quinn's *New American world*, discussed above.

drugs, herbs and flowers, grains and pulses, metals, precious stones, and colours (i.e. dyes).[54]

It is very hard to imagine Harriot not being familiar with *Divers voyages* in light of his intended journey. The structuring of information in the *Report* shares a similar approach in offering keywords, supplemented by Harriot with a discursive account based on his experience and observation. But the connection between the two men might have been more direct than Harriot simply having had sight of *Divers voyages*. They overlapped in Oxford (Harriot matriculated in 1577, the year of Hakluyt's MA) and it is possible that Harriot attended public lectures given by Hakluyt which belong to the period when Harriot was an undergraduate.[55] We can establish with certainty that they knew one another by 1586, after Harriot's return from Roanoke, when they together interviewed Nicholas Bourgoignon, a French survivor of an expedition in Florida in 1580 who had been picked up by Francis Drake.[56] It seems likely that, whatever the first date of their meeting, they became more closely associated in 1584, prior to the Virginia voyage, when Hakluyt had personal dealings with Ralegh in connection with the preparation of his manuscript presented to Elizabeth I known as the *Discourse of western planting*. Harriot was under Ralegh's patronage by that year.[57] Perhaps we should see the impulse in Harriot's *Report* to focus on the yield of information resulting from his observations and the aspiration to structure it as an indication of Hakluyt's influence.

The second feature that sets Harriot's *Report* apart from other texts printed by Hakluyt is its concern with truth. The works by Barlowe, Lane,

---

54 Richard Hakluyt, *Divers voyages touching the discouerie of America* (London, 1582), K4r-v.

55 In his dedication to Sir Francis Walsingham of *The principall navigations* (1589), Hakluyt refers to his 'publicke lectures' in Oxford demonstrating 'the new lately reformed Mappes, Globes, Spheares, and other instruments of this Art'; *Original writings and correspondence*, vol. 2, 397. For discussion of the lectures and where and in what capacity Hakluyt may have given them, see Anthony Payne, 'Richard Hakluyt's Oxford lectures', *The Journal of the Hakluyt Society* (November 2021), 1–18, online at https://www.hakluyt.com/downloadable_files/Journal/PayneHakluytOxfordLectures.pdf; Harriot's biographer, John W. Shirley, assigns the lectures to Michaelmas Term 1579 in Hakluyt's newly acquired capacity (post-MA) as 'regent master', but does not formally commit to the claim that Harriot attended them; *Thomas Harriot. A biography* (Oxford: Clarendon Press, 1983), 60. David B. Quinn sees this as probable and conjectures that Harriot had 'come under the influence of Hakluyt, who had inspired him with at least some interest in overseas voyaging'. 'Thomas Harriot and the problem of America', in *Thomas Harriot. An Elizabethan man of science*, ed. Robert Fox (Aldershot: Ashgate, 2000), 11. Another possible point of contact between Harriot and Hakluyt is referenced in the preface to *Divers voyages* (1582), where Hakluyt records a meeting with a Portuguese expert in that country's navigations that was also attended by a friend, 'a man of great skil in the Mathematikes'; *Original writings and correspondence*, vol. 1, 176. Taylor suggests that this was either Harriot or Walter Warner; *Original writings and correspondence*, vol. 1, 25 and 176, n. 2.

56 Burgoignon's relation appears in volume 3 of the *Principal navigations*. Reprinted in Quinn, *Roanoke voyages*, 763-66.

57 J.J. Roche, 'Thomas Harriot (c. 1560–1621)', in *ODNB*.

and White that I have discussed are essentially untroubled by the accusation of falsehood; they get on with the job of description without elaborate preparation or defensive remarks. No doubt this speaks to the fact that they were not intended as printed publications but began as documentary records of the journeys sponsored by Ralegh which he then made available to Hakluyt for publication. Harriot, by contrast, prepared his work for the press and fashioned his own title, though as I have indicated, the phrase 'True report' was somewhat formulaic.

Harriot's text began with an endorsement from Ralph Lane. Lane acknowledged that he might be regarded as partial (as a participant in the enterprise) and that his testimony would therefore do little to advance the acceptance of Harriot's treatise. But he insisted nonetheless that readers who approached it without a 'preiudicate minde' would find 'true enformation' (319). He offered his own credit to back up the claim: 'things universally are so truly set downe in this treatise by the author thereof, an Actor in the Colony, or a man no lesse for his honesty then learning commendable'. Lane went so far as to say that the work ranked 'euen amongst the most true relations of this age' (319).[58]

When Harriot commences the text it becomes clear why Lane's testimonial was necessary. Harriot alluded to 'slanderous and shamefull speeches' that had been 'bruited abroade' (320), evidently by mariners who had returned from the voyage, particularly those who accompanied Sir Richard Grenville back to England in 1585. It becomes obvious that a battle for public opinion was taking place, and Harriot wrote to shore up those 'Favourers and Welwillers' of the colonial project, as he called them. He drew on his authority as one experienced in the discovery of the territory, with special responsibility for dealing with the natural inhabitants, in order to demonstrate the falsity of the accusations and slanders. The Blatant Beast was evidently at work.[59] Harriot acknowledged at the same time that divergent relations of Virginia had circulated, leaving many in doubt, a fact that required some explanation. First he attacked the source of the aspersions, noting that some who spread these ill reports had been taken to task for misdeeds and speaking ill of their governors. Others were essentially ignorant of Virginia; but being back home among friends where no one could gainsay them, they built themselves up as knowledgeable figures, 'and make no men so great trauailers as themselues' (322). Having

---

58 Quinn points out (*Roanoke voyages*, 320, n. 1) that de Bry omits Lane's letter and inserts his own epistles; see *Roanoke voyages*, 399–402 for these. For a wider discussion of credit, see Daniel Carey, 'The problem of credibility in early modern travel', *Renaissance studies*, 33, no. 4 (2019), 524–47.

59 This is Edmund Spenser's allegorical figure in *The Faerie Queene*, possessed of various unpleasant attributes, including slander, evident when he pours his 'poysnous gall forth to infest/The noblest wights with notable defame'; *The Faerie Queene*, ed. A. C. Hamilton (London: Longman, 1977), VI.vi.12.5-6 (661).

been there long enough they had to maintain that they knew what was true and false:

> Of which some have spoken of more then ever they saw or otherwise know to bee there, othersome have not bin ashamed to make absolute denial of that which ... by others is most certainly and there plentifully knowne. (323)

Having diagnosed the type, he needed to say more to undermine their credibility. He attributed their ignorance to several factors. Some of them had never travelled outside the island where the colony was seated and therefore knew little of the situation. Others, finding no gold and silver in Virginia, cared for nothing but to pamper their bellies; others still, of a 'nice bringing vp' (323) in cities or towns, fell to disparaging the place because it did not contain fair houses or cities and lacked familiar dainty foods and beds of down and feathers; they were too soft in other words for the task in hand.

The issue of trustworthiness bedevilled early modern travel writing; the carping between André Thevet and his great rival Jean de Léry over their time in Brazil constitutes the great precedent for what Harriot worried about.[60] But he does more than merely tackle the issue in his prefatory remarks. Throughout the text, we can see the evident care he took not to overextend his claims, at times even sharing with the reader disappointing results of his inquiries and assessments. Thus he shores up his reliability by not overstating the case, in spite of his overt desire to attract support for the venture. In doing so he avoids the heady prose of Barlowe or the exuberant tone of Lane's 1585 letter to Hakluyt.

The final feature I would like to call attention to in Harriot's *Report* is the frequent refrain of trial, proof, and experience that runs throughout it. This emphasis complements the task of establishing and reinforcing his credibility. But it also relates to Harriot's formation in natural philosophy, and changing canons of experiment and evidence in which he participated. No other document in the Virginia sequence published by Hakluyt observes the same care and attention to such matters.

Several examples can be mentioned out of a substantial number that populate the text. Early on, Harriot speaks of the wood of sassafras and notes: 'It is found by experience to bee far better and of more uses then the wood which is called Guaiacum, or Lignum vitae' (329).[61] Under his heading for iron, he notes that the 'minerall man' (Joachim Gans or Ganz, most likely) made trial

---

60 See Frank Lestringant, *Le Huguenot et le sauvage. L'Amérique et la controverse coloniale en France, au temps des guerres de religion (1555-1589)*, 3rd edn. (Geneva: Droz, 2004).

61 He referred the reader here to Nicholas Monardes, *Ioyfull newes out of the newe founde worlde*, trans. John Frampton (London, 1577). Monardes discusses sassafras at M2r-O4v.

of rocks there and found them to 'holde yron richly' (331).[62] The trial of copper also determined that they contained silver ore. Potentially valuable dyestuffs existed, including the bark of a local tree that produced a sort of red. Nonetheless Harriot remarked that 'their goodnesse for our English clothes remayne yet to be proved' (335). He expressed disappointment that although they arrived with sugar canes, the season had passed to plant them and they 'could not make the proofe of them as wee desired' (336). His estimate of crop yields was made 'according to the rate by us experimented' (342),[63] though of wheat they 'could make no triall' (344); similarly, with the valuable dyestuff cochineal – whose origins were still uncertain – he affirmed that it was not a fruit but found on the leaves of a plant, but what leaves he admitted 'we have not so specially observed' (352).[64] (Hakluyt helped out with a marginal note but did not recognize the source as an insect.[65]) Harriot's qualifications in the domain of science and mathematics are well known; my point is that they are manifested in the text by establishing an experimental relationship to knowledge. This runs all the way from his calculations on the productivity of land to less congenial matters of diet which were subject to experience, including wolves and dogs on whom members of their party had 'experimented' (357).

Harriot's manner of organizing his text and reportage undoubtedly places him closer to the Hakluyt end of the spectrum in privileging information, predicating the task on documenting knowledge, engaging in careful observation, and accumulating a grounded understanding of resources and the potential for profitable settlement. Framing this discussion within an optimistic assessment was also clearly consistent with Hakluyt's sensibility as a promoter of overseas endeavour. Small wonder that he assisted with making it better known by collaborating to produce the de Bry edition with such lasting influence.

62 On Gans (or Ganz) and 'knowledge-making' more generally in Harriot's text, see Stephen Clucas, 'Thomas Harriot's *A briefe and true report*. Knowledge-making and the Roanoke voyage', in *European visions. American voices*, ed. Kim Sloan (London: The British Museum, 2009), 17–23 (on Gans, 18).

63 Harriot continued: 'I can assure you that according to the rate we haue made proofe of, one man may prepare and husband so much grounde (hauing once borne corne before) with lesse then foure and twentie hours labour, as shall yield him victuall in a large proportion for a tweluemoneth'; Quinn, *Roanoke voyages*, 343. On this point, see B.J. Sokol, 'The problem of assessing Thomas Harriot's *A briefe and true report* of his discoveries in North America', *Annals of science*, 51, no. 1 (1994), 13.

64 Harriot pointed in the right direction by identifying the prickly pear (under the indigenous name *Metaquesúnnauk*) as connected with cochineal (the cactus leaves attract the insect from which the dye is derived). But he acknowledged that whether 'it be the true cochinile or a bastard or wilde kinde, it cannot yet be certified' (352). On cochineal, see Amy Butler Greenfield, *A perfect red. Empire, espionage, and the quest for the color of desire* (New York: HarperCollins, 2005).

65 Hakluyt's note to this passage in the second edition of *The principal navigations*, vol. 3, 273, observes that 'There are iii kinds of Tunas [i.e. prickly pears, see OED] whereof that which beareth no fruit bringeth forth the Cochinillo'.

Yet we must also confront the disjunctions made apparent when we read this work alongside Lane's narrative of conflict. There is no head in Nugent's hand, to borrow the title of Michael Leroy Oberg's book, no treacherous Pemisapan to contend with. He appears in Harriot's *Report* but in a benign guise, and notably under the name Wingina, not the more hostile one of Pemisapan.[66] Harriot's relatively sanguine view of the local inhabitants as harmless, due to their lack of dangerous offensive weapons,[67] is undermined by learning of the considerable skill exercised in their attacks that we later discover in White's account.

If we are to place the *Briefe and true report* in a continuum of the texts printed by Hakluyt, it does not share the exculpatory objectives of Lane or White. But it does demonstrate the same concern for reputation, in Harriot's case registered in the need to defend his truthfulness and integrity against slander. The journal form, eschewed by Harriot, had the benefit of straightforwardness but it often produced narrative lacunae, as we have seen. Lane's work, designated as a 'discourse', shows how difficult such a form was to manage coherently while relating historical incident. Harriot is unquestionably the most accomplished of the four authors in controlling a piece of writing, giving it a consistent shape and conclusion. Barlowe's account is perhaps the closest in focus and purpose to the *Report*. Comparing the two makes us realize just how successfully Harriot has transformed experience and observation into information.

Over a 12-year period, Harriot's *Briefe and true report* appeared on four occasions: first, the London quarto of 1588; the second, in a run of sources on Virginia published by Hakluyt in 1589 in *The principall navigations*; the third, in the de Bry folio edition of 1590 accompanied by John White's re-engraved plates of indigenous scenes and people; and the fourth, in the final volume (1600) of *The principal navigations* where White's narrative of his final journey concluded the sequence. My point in this discussion has been to emphasize what we can learn by studying Harriot's text as Hakluyt presented it. Not only do the surrounding texts printed by Hakluyt inflect the meaning of what Harriot imparted, they enable us to appreciate the distinctiveness of Harriot's approach and the sensibility that informed it. Reading Harriot's *Report* as an embedded text, alongside Barlowe, Lane, and White, provides us with a different perspective, not only on the range of rhetorical modes available to these authors but also on the conditions of

66 Harriot notes that Wingina joined them in prayer at times and that on two occasions when Wingina was grievously ill and fearing death he asked the English for their prayers and to intercede with the English God on his behalf. Seeing the exceptional death rate in towns where the 'naturall inhabitants' had engaged in 'wicked practises' against the English, Wingina concluded that his illness was the work of the English God; Quinn, *Roanoke voyages*, 377-79.

67 Quinn, *Roanoke voyages*, 369.

knowledge and encounter, which depended on assessments of character and interpretation of signs as much as on empirical survey.

By the time Hakluyt reprinted Harriot's work for a final time in 1600, the trajectory of Virginia's story could only really be taken as one of failure and disappointment, placing it closer to the French experience in Florida in the 1560s than to the Spanish one in the New World. If so, the message may have been the one Hakluyt highlighted in his dedication to Sir Walter Ralegh of his translation of René de Laudonnière's *Histoire notable de la Floride* in 1587. There Hakluyt 'forewarned and admonished' of the 'grosse negligence in providing sufficiencie of victuals, the security, disorders, and mutinees that fell out' during the French venture. The point was for Ralegh and others to 'learne to prevent and avoyde the like' while also 'reading of the manifolde commodities & great fertilitie' of a territory with 'affinitie, resemblance or conformitie with' Virginia.[68]

Nor should we overlook the significance for contemporaries of reading Harriot in the context in which Hakluyt presented him, that is, as someone immersed in a shared narrative emplotment rather than as author of a self-contained account. Captain John Smith's *Generall historie of Virginia, New-England, and the Summer Isles* (1624) draws closely – and overtly – on volume 3 of *The principal navigations* by retelling Barlowe, Lane, Harriot, and White in sequence.[69] All of the key actors and scenes are here, from the beating surf replete with grapes found in Barlowe's account to the fruit-fulness of the soil; Lane's accumulation of incident, including his reliance on the assistance of Ensenore and Skiko through to the treachery of Pemisapan and his beheading by Lane's men. This section is succeeded by Harriot's text. Smith retains the structure and emphasis of the *Briefe and true report* by starting with 'marchandize and Victualls' before moving on to the people and their manners. Next in the sequence comes Sir Richard Grenville's journey to relieve the colony, only to find it abandoned.[70] Smith ends with the two final voyages of John White, in 1587 and 1590, remarking in his own words, tersely, 'And this was the conclusion of this Plantation, after so much time, labour and charge consumed.'[71]

---

68 *Original writings and correspondence*, 372.
69 For the acknowledgement of Hakluyt, see John Smith, *The generall historie of Virginia, New-England, and the Summer Isles* (London, 1624), 2.
70 The group he left behind grows, in Smith's rendition, from 15 to 50. See his *Generall historie*, 13.
71 Smith, *Generall historie*, 16. Valuable annotations on Smith's use and departure from Hakluyt are provided in Philip L. Barbour's critical edition. See *The complete works of Captain John Smith (1580-1631)*, ed. Philip L. Barbour, 3 vols. (Chapel Hill: University of North Carolina Press, 1986), vol. 2, 61-88.

6 'Cause Both to Feare and Love Us': How to Found an Empire in Harriot's Day

Felipe Fernández-Armesto

At the risk of uttering an obvious *captatio benevolentiae*, I am tempted to call my subject 'the less interesting Thomas Harriot' – or at least, the aspect of his mainly twofold career that has attracted less interest. Harriot the imperialist, though by no means ignored in scholarship, has inspired only a small minority of previous lecturers in the present series, whereas his scientific work enormously preponderates in the literature and constitutes the foundation of his reputation as a luminary and perhaps a beacon of early modern enlightenment.[1] The dichotomy is in some ways misleading: Harriot's interest in navigation overlapped astronomy and mathematics on the one hand and the practical requisites of overseas empire-building on the other. He took part in strenuous adventures in Ireland and Virginia with Sir Walter Ralegh primarily because, according to his own protestations, he needed patrons and wanted opportunities to study.[2] His imperial or proto-imperial ventures in those countries furnished opportunities for observations relating to ethnography, botany, geography, and climate. His expertise in surveying colonizable land was part of his well-stocked locker in mathematics. His fame as a scientist is, in part at least, an unintended consequence of his dabblings in empire: the only work of his published in his lifetime was

---

1 In the first lecture in 1990, on 'Thomas Harriot and the problem of America', published in R. Fox (ed.), *Thomas Harriot. An Elizabethan man of science* (Aldershot and Burlington, VT, 2000), 9–27, David Quinn set an example unfollowed until Surekha Davies gave 'Thomas Harriot, John White, and the invention of the Algonquian Indian' in 2011, since when the imperial theme has preponderated in lectures by Lesley Cormack, 'The Whole Earth: a Present for a Prince' (2012); David Reed Sacks, 'The true and certain discovery of the world: Thomas Harriot and Richard Hakluyt' (2014); Mark Horton, 'Thomas Harriot: the world's first ethnographer?' (2017); Daniel Carey, 'Harriot, Hakluyt and new found Virginia' (2018); and Larry E. Tise, 'Thomas Harriot and the creation of America's first illustrated coloring book' (2021).

2 'never any busy meddler in affairs of state, never ambitious for preferments, but contented with a private life for the love of learning that I might study freely'. J. W. Shirley, *Thomas Harriot. Renaissance scientist* (Oxford, 1974), 29; G. Batho, 'Thomas Harriot and the Northumberland household', in Fox (ed.), *Thomas Harriot. An Elizabethan man of science*, 28–47.

DOI: 10.4324/9781003096580-7

A briefe and true report of the new found land of Virginia, a promotional pamphlet designed to refute critics of the Virginia enterprise and attract investors and settlers to England's first, doomed American outpost, where Harriot accompanied an abortive attempt at colonization in 1585. Had Richard Hakluyt, the restless promoter of a vision of English empire, not encouraged its publication, and had a spectacular illustrated edition not followed in Frankfurt, scholars might have overlooked Harriot altogether, and his gifts for mathematics, physics and astronomy might have been forgotten.

The emphasis and interest, however, of the *Briefe and true report* is on a problem that engaged and usually defeated the author and many of his contemporaries: how to found an empire, at a great distance from the metropolis, in an alien and apparently intractable cultural environment, in the unpropitious circumstances of a pre-industrial world.

In Harriot's day, the obvious model to follow – the only example of conspicuous success – was that of the Spanish global monarchy. Spaniards made Romans their model empire-builders because they had no other prototype to follow.[3] The English had the advantage of being able to copy the Spaniards. It made sense to do so. Even before absorbing Portugal's dominions in 1580, Philip II's monarchy stretched from Manila to Milan and from Chile to Chihuahua and the Chesapeake. It criss-crossed more of the world than any predecessor and comprised a greater diversity of biomes and cultures. The rapidity with which it arose baffled its beneficiaries, who often ascribed it to providence or miracles. The projectors of an English or, as they sometimes said, 'British' empire, who included prominent members of the circles of patronage and study that surrounded Harriot, such as Ralegh, Gilbert, and Hakluyt, gathered all the intelligence they could obtain on the Spaniards' proceedings and devoured all the relevant writings to which they gained access. Harriot made their envy explicit in the *Briefe and true report* in apologizing to his readers for Virginia's lack of precious metals and apparently limited economic potential. He pointed out, perhaps disingenuously, that the Spanish empire began relatively modestly compared with the value of the acquisitions Spain went on to make, and that the prospects for extending an English colony into more profitable regions therefore beckoned. 'Why may wee', he remarked, 'not then looke for in good hope from the inner parts of more and greater plentie ...? Unto the Spaniardes happened the like in discovering the maine of the West Indies'.[4]

---

3 S. MacCormack, *On the wings of time. Rome, the Incas, and Spain* (Princeton, NJ, 2007), 5–12, 274; A. Grafton, *New worlds, ancient texts. The power of tradition and the shock of discovery* (Cambridge, MA, 1995); J. H. Elliott, *The Old World and the New. 1492–1650* (Cambridge, 1992).

4 D.B. Quinn (ed.), *The Roanoke voyages*, vol. 1 (London, 1952), 383.

Envy is a sincere form of flattery and English projectors strove to turn their envy of the Spanish empire, by way of imitation, into emulation. Queen Elizabeth's hand, laid on South America in the so-called Armada portrait, sums up the ambitions of Gilbert, Hakluyt, Ralegh, Dee, and other proponents of English out-thrust across the Ocean.[5] Richard Eden translated a Spanish-based humanist's account of the founding of the Spanish New World and Martín Cortés's navigational manual on how to get to it.[6] Richard Hakluyt was explicit about the need to imitate the Spanish model in his *Discourse of western planting*.[7] When, after the failure of the enterprise in which Harriot engaged, the English returned to Virginia in the new century, Captain John Smith's programme was clearly set out in the engravings that illustrated the history of his deeds: he shows himself repeating the strategy of Cortés and Pizarro, capturing a native paramount and making a puppet ruler of him.[8]

The *Briefe and true report,* I hope to suggest, demonstrates the author's indebtedness to Spanish predecessors. Though much cited and, presumably, often read, even – in some curricula – in extracts by schoolchildren,[9] the document has been the object of surprisingly little serious study; it occupies, for instance, only seven pages in a recent substantial biography of Harriot.[10] The value of examining passages that recall Spanish imperial literature and experience is not only that they confirm English dependence on, or at least reference to, Spanish precedents, but also that they provide clues towards the solution of a further problem in the history of English imperialism: that of why it started so badly.

In some ways, the early failures of England's experiments in transatlantic exploration and colonization are surprising. The country had the advantage of an Atlantic-side position, privileged access to the trade winds that linked

5 J. Hart, *Representing the New World. The English and French uses of the examples of Spain* (New York, 2000), esp. 94–116; F. Fernández-Armesto, 'Inglaterra y el Atlántico en la Baja Edad Media', in A. Bethencourt Massieu (ed.), *Canarias e Inglaterra a través de los siglos* (Las Palmas, 1995), 11–28; J.H. Elliott, *Empires of the Atlantic world. Britain and Spain in North America* (New Haven, CT, 2006), xviii.

6 *The Decades of the newe worlde or weft India conteynyng the nauigations and conqueftes of the Spanyardes* (London, 1555); *The Arte of nauigation conteyning a compendious description of the sphere, with the making of certayne instruments and rules for nauigations, and exemplifyed by many demonstrations. Written by Martin Cortes Spanyarde. Englished out of Spanishe by Richard Eden* (London, 1561).

7 R. Hakluyt, *Discourse of western planting,* ed. D.B. and A.M. Quinn (London, 1993), 115–16, 119–20.

8 *The generall historie of Virginia, New England, and the Summer isles. With the names of the adventurers, planters, and governours from their first beginning, Ano: 1584, to this present 1624 (* London, 1624).

9 https://www.encyclopedia.com/people/science-and-technology/mathematics-biographies/ thomas-harriot, consulted 11 June 2020.

10 R. Arianrhod, *Thomas Harriot. A life in science* (Oxford, 2019), 80–82, 84–87, 89, 102–3.

Europe and Africa to the New World, relevant prior experience in Ireland (and to some extent in internal colonization within Great Britain, comparable to Spain's and Portugal's in the so-called Reconquest),[11] a large and professional merchant marine, access to all the requisite technical knowledge, adequate manpower for the job, and merchants who were interested in growing their trades and endowed with plenty of spare capital to invest in overseas ventures. There is no obvious and determinant reason why England's should have been a laggard empire in comparison with those of Castile or Portugal.[12] The *Briefe and true report* shows one of the ways in which English empire-builders erred and delayed the effective realization of their hopes: Harriot misread some of the Spanish evidence and underestimated the obstacles that faced him and his fellows. Before reviewing the evidence, we need to gauge the nature of the undertaking in which Harriot and his fellow projectors, propagandists and venturers engaged, and the extent of the difficulties they faced.

'He told us an empire is like a duck', said an old student, recalling one of my lectures at a class reunion. 'None of us will ever forget the image but I, for one, must confess that the reason for the analogy eludes my memory'. The resemblance would be unmemorable if it were obvious, but an empire is like the duck in a famous anecdote, told, like all good anecdotes, of various individuals but always associated with McCarthyism. In reply to the journalist's question, 'How can you tell who is a communist?', the witch-hunter replies, 'If it looks like a duck, walks like a duck and quacks like a duck, it's a duck'.[13] In other words, what is hard to define – be it communist, duck, or empire – may be characterized by profiling. 'Empire' is an elusive term. A few years ago, Leonard Blussé, the late Chris Bayly, David Armitage, Shruti Kapila, and I were dining together in Cambridge, Massachusetts. Naturally we discussed empires and, equally naturally, disagreed about almost everything. One matter on which we concurred, however, was that none of us could say what an empire is – a chilling finding, as it suggested that five supposedly great experts on the subject literally did not know what we were talking about. Recent attempts at definition from historians and political scientists justify our disavowal.[14] At the time of Harriot's Virginian experience, there were at least thirty states in the world that historians commonly refer to as empires. Yet they had nothing in common that collectively distinguished them from other kinds of polity. They ranged from

---

11 A.L. Rowse, *The expansion of Elizabethan England* (New York, n.d. [1955]), 1–157.

12 F. Fernández-Armesto, 'New worlds', in J. Davey (ed.), *Tudor and Stuart seafarers* (London, 2018), 21–36.

13 E. R. Bayley, *Joe McCarthy and the press* (Madison, WI, 1981), 129.

14 P. Pomper, 'The history and theory of empires', *History and theory,* 44 (2005), 1–27; F. Cooper and J. Burbank, *Empires in world history* (Princeton, NJ, 2010), 2, 6, 8–13; D. Day, *Conquest* (Oxford, 2012), 4–7. Cf. S. N. Eisenstadt, *The political systems of empires* (New Brunswick, 1963), 10–12.

loosely articulated dynastic agglomerations, such as the Omani empire in and around the Indian Ocean, to tribute-levying rackets such as Russia's in Siberia, and vast conquest-states such as the Mughal Empire. That of the Ming had a uniform system of government, whereas most, like those of the Ottomans, say, or Safavids, were patchworks of different and sometimes conflicting systems, jurisdictions, and institutions. Almost all, in common with most states, were forged in conquest. But 'negotiated obedience'[15] – to borrow a term from Jesuit devotional and missionary practice – also played a part in the formation of empires, and 'tu, felix Austria, nube'. Some empires formed inland, of contiguous territories acquired with the aim of mastering population or products; others were 'seaboard empires', linking ports and routes in order to control particular trades.[16] Some, like those of the Uzbeks or Altan Khan, were in the tradition of 'the empire of the steppes', whereas most were agrarian.[17] All were large in relation to states that preceded or followed them in their own territories, but they varied in size from the compact, in south-east Asia, to the vast and continuous, like Russia's, or the globally dispersed, like Spain's. Like most polities, almost all were ethnically and culturally diverse, but some, as in Japan or Vietnam, were practically homogeneous. In many, rulers or elites embraced a sense of universal mission, such as conversion to a religion or extension of the Mandate of Heaven, but others practised mere opportunistic land-grabbing.

In the absence of common, distinguishing hallmarks that might support a definition, the best way to understand empires is to resort to the McCarthyite strategy and compose a profile of a state of an appropriate kind, in awareness that no empire will meet all the typical criteria. When I invite students to make such a list, predictable ingredients usually include size, diversity, formation by conquest, and relationships of subjection and victimization. The commonest ingredient students propose is power, presumably because it is hard to see why, except by force, people across vast swathes of territory, with their own cultures and political traditions, should accept or endure immersion, along-side communities with whom they otherwise share little or nothing, under a common, alien focus of allegiance. Yet it is impossible to understand empires, especially those of pre-industrial eras, without acknowledging their funda-mental and pervasive weakness: the lack of adequate means of communicating or enforcing commands. In the early modern period, even empires that laid heavy hands on subjects at their centres stretched feeble fingertips towards their peripheries. When Jerónimo de Quadros, Portuguese commander of a

---

15 F. Alferi and C. Ferlan (eds.), *Avventure dell'obbedienza nella Compagnia di Gesù. Teorie e prassi fra XVI e XIX secolo* (Bologna, 2012), 197.

16 C. Verlinden, *Koloniale expansie in de 15de en 16de eeuw* (Bussum: Fibula-Van Dishoeck, 1975). Or 'seaborne' empires in a phrase that served as part of the title of each in a series of books written or edited by J.H. Parry.

17 R. Grousset, *The empire of the Steppes. A history of central Asia* (New Brunswick, 1970).

fort on the coast of what is now Iran, described his difficulties to his king, Philip II of Spain, in 1588, the year of publication of Harriot's *Report*, he listed deficiencies of time, cash, fortifications, men, and munitions (that is, of arrows, for firearms were utterly unavailable); his most deeply felt want, however, was of adequate supplies of opium to feed the men of his garrison, whose task was so dispiritingly hopeless that they could only face it with the aid of narcotics. I know of no more graphic measure of the weakness, at its edges, of what was universally acknowledged as one of the most powerful empires of the day. The captain's pitiful submission never even reached its destination, being captured en route by English pirates and ending up in the National Archives of the United Kingdom: communications were as precarious as the fabric of Quadros's mud-built fort, which he 'had to rebuild every year after the rains'.[18]

Even with the crushing resources of technology and manpower at the command of states today, the effectiveness of imperial control over subject populations is unpredictable, and resistance is often effective, as the collapse of most European overseas empires in the third quarter of the twentieth century and the experiences of intruders in Iraq, Afghanistan, and Vietnam has shown. The only sure way to counter resistance is to head it off by securing the compliance of a critical mass of indigenous supporters, preferably among incumbent elites who command followers of their own, in subject territories. Successful empires – those that last and prosper – are necessarily collaborative ventures, in which interdependent elites located at the centre and the periphery, in the heartland and the acquired territories, share the advantages for economy and security that a *Grossraum* confers.

Thomas Harriot knew the right formula: a strategy, as it were, of stick and carrot. From the point of view of imperial powers and colonizing communities, Harriot acknowledged, it is necessary for indigenous people in acquired territories 'both to feare and love us that shall inhabite with them'.[19] Fear and love were, to Harriot and his contemporaries, obviously compatible and even complementary qualities, commanded of worshippers in respect of God and children in respect of parents.[20] Columbus expressed hopes that the inhabitants of Hispaniola would submit to Spaniards 'out of fear and love'.[21] Francisco de Vitoria allowed, with some reservations, the presumed child-like innocence of Native Americans as a means of legitimizing Spanish rule over them on the strength of an

18 F. Fernández-Armesto, *Millennium* (London, 1995), 228.

19 Quinn (ed.), *Roanoke voyages*, vol. 1, 368.

20 L. Stone, *Family, sex and marriage in Tudor England* (London, 1977), 99–101; H. A. Enno van Gelder, *The Two Reformations in the sixteenth century* (The Hague, 1961), 331; G. Strauss, *Luther's house of learning* (Baltimore, MD, 1978), ch. 8; P.F. Grendler, 'The schools of Christian doctrine in sixteenth-century Italy', *Church history*, 53 (1984), 319-31 (329).

21 F. Fernández-Armesto, *Columbus on himself* (Indianapolis, 2010), 91.

analogy with a paterfamilias.[22] For the applicability of love and fear in empire-building and for means of harnessing them, Harriot could draw on Spanish historical experience and some of the texts that recorded it. The inferences he made included five in particular, revealed in the *Briefe and true report*: that conquest would be easy; that indigenous superstitions would facilitate it; that the invaders' technological superiority would assist it; that divine Providence would guarantee it; that religious influence would confirm it; that disease would undermine resistance; and that the Bible would impact upon it. Harriot read his and his fellow-expeditionaries' experience as affirming these inferences. Yet, in my submission, all were in some measure delusive. In palliation of Harriot's errors it can be said that most historians have shared them or have been misled in similar ways.

The speed with which Spaniards acquired vast territories and spectacular riches in America is at the core of the delusion that New World empires were easy to conquer, primarily because of the military superiority of the invaders – superiority compounded of superior prowess allied to superior technology. At first glance, the facts seem consistent with the myth. An impartial observer, beholding the world in the first couple of decades of the sixteenth century might have been impressed by the rate of Ottoman or Muscovite success but would surely have noticed that among the four most impressive cases of rapid imperial expansion were those of the people we usually call Aztecs and Incas. Yet Spaniards seemed to gobble up both their empires almost at a gulp. Cortés was master of Tenochtitlan within four years of his arrival on the American mainland. It took Pizarro little more time to dominate Tawantinsuyú. The rapidly unfolding narratives were familiar in Europe to readers of Cortés's own printed reports by 1524 and in the case of Peru, thanks to that of Francisco Xérez, published in 1534, and its successors. The following year, volumes of Gonzalo Fernández de Oviedo's comprehensive account of Spain's American conquests echoed earlier works in exalting Spaniards' heroism and confirmed readers' sense that empire had been accomplished with ease.[23] Harriot expected the English to be similarly favoured in Virginia. Conquest, with care and prudence properly exercised, would come easily. 'By carefulnesse of ourselves neede nothing at all to be feared ... expecting the goode success of the action'.[24]

The problem of how Spain's empire arose so easily has been the main focus of the historiographical tradition, at least since Francisco Xérez. Nowadays we tend to stress previously underestimated delays and difficulties. The last

---

22 B. Premo, *Children of the Father King. Youth, authority, and legal minority in colonial Lima* (Chapel Hill, NC, 2005), 32–34; F. de Vitoria, *Political writings*, ed. J. Laurance and A. Pagden (Cambridge, 1991), 240–91.

23 Roberto Gonzalez Echevarría and Enrique Pupo-Walker (eds.), *Cambridge history of Latin American literature*, vol. 1 (Cambridge, 1996), 119.

24 Quinn (ed.), *Roanoke voyages*, vol. 1, 382, 387.

Inca stronghold in Vilcabamba was unmolested until 1572. In what had been the Aztec zone of preponderance, indigenous elites remained in local and regional control of their communities, yielding autonomy and land only very gradually, throughout the colonial era. The last Maya kingdom did not end resistance and submit to Spain until 1697. It took until 1793 for Spanish negotiators to make suitable arrangements with the Mapuche. Remote and intractable environments never experienced more than nominal Spanish rule. Increasingly – though not yet sufficiently – scholars are showing awareness that Spanish 'conquests' were typically conquests of some indigenous communities by others,[25] with Spaniards benefiting for so far poorly understood reasons, on which, I hope, the present essay will cast some light. Indeed, the notion that Spain's Native American subjects were easy to conquer arose not so much from the facts, as they happened, as from falsehoods people believed.

Evidence of the conquistadores' superiority derives largely from their own accounts, which were, of course, distorted by self-interest. Like graduates writing 'personal statements' in pursuit of employment today, conquistador-raconteurs found modesty an encumbrance and accuracy a superfluity. The object of writing a *probanza de méritos* – a document, in which most conquistador narrations originated, in solicitation of rewards from the crown for services rendered – was not to tell the truth but to make a favourable impression. Examined severally, many of those that survive are frankly incredible. To make an example of one, which was accessible to Harriot in Fernández de Oviedo's book, we may turn to the account of the exploits of Diego de Salazar, conquistador of Puerto Rico.[26] If the document were believable, we should have to acknowledge that while natives were feasting in preparation for the ritual sacrifice of a woefully disheartened captive Spaniard, the hero crept unseen into the heart of a hostile village to rescue him and lead him, single-handedly through the midst of the foe, fighting off three hundred warriors, fatally wounding their chief and escaping, bloody but unbowed. As if this feat were insufficient, Salazar returned to the enemy camp in answer to a summons from the moribund chief. His companion begged him on bended knees to refuse the request, but Diego demurred: he would rather return and die than flee in fear. Far from exacting revenge, the affronted chief begged to be allowed to take Salazar's name. Salazar assented and went away laden with gifts of jewels, food, and slaves. Thereafter, 'I fear you as if you were Salazar' became a proverbial testimonial among the natives. Such sources reveal not only the self-congratulatory agenda that misled credulous readers into believing Spaniards' accounts of their own prowess, but also a further source of distortion: the influence of the romances of chivalry that were conquistadores' most favoured reading matter and shaped the narrators'

25 J.E. Kicza, *The Indian in Latin American history* (Lanham, MD, 2004); M. Restall, *Maya conquistador* (Boston, MA, 1998).
26 F. Fernández-Armesto, *Our America* (New York, 2014), 21–22.

images of themselves. Indeed, Fernández de Oviedo had written a work in the genre himself before taking up his role as historian. The image of Spanish superiority in war found its way into the historical tradition not because it is reliable but because it served the narrators' ends and reflected the literary influences to which they were susceptible.

Alongside superior prowess, Europeans' supposed technological superiority was a topos of the early literature of conquest, not only in the Americas. Harriot echoed it and claimed that experience confirmed it in the case of the English in Virginia. '... [W]hat their fight is likely to be, we having advantages against them so many manner of ways, as by our discipline, our strange weapons and devises else; especially by ordinances great and small, it may be easily imagined; by the experience we have had in some places, the turning up of their heels against us in running away was their best defence'.[27] Some historians have remained convinced of the utility of appealing, when trying to understand what happened in early modern encounters between natives and Europeans, to the supposed invincibility of the techniques and technologies invaders brought from Europe.[28] In some ways, the technology gap was indeed decisive. Harriot praised native boatbuilding in Virginia for local purposes,[29] but there is no scope to question the importance of European nautical technology in getting the invaders across the Atlantic in the first place and in crucial episodes of riverine and lacustrine warfare in the Americas.[30] One can see, for instance, in the Lienzo de Tlaxcala, a source compiled by indigenous reporters of the fall of Tenochtitlan, the brigantines that Spanish craftsmen extemporized for the siege, dominating Lake Texcoco. Other Spanish technical achievements that impressed natives appear in the same picture: a monstrously conspicuous suit of armour; the steel edges of Spanish blades; the little cannon, mounted on board the boats, such as native porters toiled to haul from the coast to add to the besiegers' armoury.[31] It is doubtful, however, whether all this panoply, together with the horses and dogs that feature prominently in many historians' perceptions, made much, if any, practical difference. Inaccuracy, unreliability, and the difficulty of importing shot reduced firearms to marginal significance. At Tenochtitlan, Cortés had only a handful of firearms of all descriptions. Powder was too volatile for many New World environments. Steel weapons were few and rarely wielded, because, on the

27 Quinn (ed.), *Roanoke voyages*, vol. 1, 371.
28 C. Townsend, 'Burying the white gods: new perspectives on the conquest of Mexico', *American historical review*, 108 (2003), 659–87; P.T. Hoffman, *Why did Europe conquer the world?* (Princeton, NJ, 2015).
29 Quinn (ed.), *Roanoke voyages*, vol. 1, 363-64.
30 F. Fernández-Armesto, 'Naval warfare after the Viking age', in M. Keene (ed.), *Medieval warfare. A history* (Oxford, 1999), 230-52.
31 http://www.latinamericanstudies.org/tlaxcala/Tlaxcala-lienzo-10.jpg, consulted 3 August 2020.

whole, Spanish participants in battle had the good sense to anticipate the methods of the Duke of Plaza-Toro, who 'led his regiment from behind; he found it less exciting'. Horses were useful in some terrains but again usually few in number and easily captured and mastered by adversaries. Technological advantages of all kinds rarely have longstanding impact on the course of wars because of such appropriations and because familiarity rapidly erodes their psychological impact.[32] Harriot was aware of how much his Native American acquaintances admired his scientific gadgets and can be pardoned for over-estimating the difference war-technology might make in Virginia. Spanish precedents, however, supported only very partially his expectation of great consequences from 'strange weapons and devises'.[33]

Superior technology was one aspect of a larger advantage from which Harriot anticipated great results: superior science, manifest partly in in-digenous peoples' awe of intruders' claims to knowledge of and power over nature, and partly in the superstitions that subverted their resistance. Harriot, by his own account, entranced Native American audiences with tricks of apparent prestidigitation with, among other gadgets, clockwork, magnets, magnifying glasses, and what he calls 'a perspective glass whereby was showed many strange sights'.[34] He makes no claim for these as in-struments of peaceful conquest, but they contribute to his picture of awe-struck ingénus, implicitly acknowledging the newcomers' mastery. The assumption that omens cowed natives was already an indelible element in Spanish traditions. Harriot echoed it when he declared that inhabitants of Virginia were predisposed to demoralization by 'knowing of the Eclipse of the Sun which we saw the same year before in our voyage thitherward, which unto them appeared very terrible'.[35] A 'Comet which began to ap-peare but a few days before the beginning' of the plague that accompanied the invaders had, in Harriot's estimation, a similar effect. Eclipses were routine to astronomers and hardly belonged among the 'great mutations in the sun and moon' that educated Europeans might mistake for celestial advertisements. They held, of course, no terrors for Harriot. His ob-servations of the eclipse on the outward voyage to Virginia, at sunset on 19 April 1585, have attracted praise for their accuracy from modern in-vestigators.[36] He was well acquainted with comets, and was said by a contemporary to have studied all comets observable in Europe in his

32 D. R. Headrick, *Power over peoples. Technology, environments, and Western imperialism, 1400 to the present* (Princeton, NJ, 2010); M. Restall, *Seven myths of the Spanish conquest* (Oxford, 2003), 14–44.

33 Quinn (ed.), *Roanoke voyages*, vol. 1, 371.

34 Ibid., vol. 1, 375.

35 Ibid., vol. 1, 380–81.

36 J.J. Roche, 'Thomas Harriot's observations of Halley's comet in 1607', in J.D. North and J.J. Roche (eds.), *The light of nature. Essays in the history and philosophy of science presented to A. C. Crombie* (Dordrecht, 1985), 175–92.

lifetime, though no detailed observations of the event he saw in Virginia seem to have survived.

There is no reason to suppose that he thought eclipses or comets could affect or predict terrestrial events, but he evidently supposed that Virginia's inhabitants did – not necessarily because the supposition was valid but rather, perhaps, because it was a commonplace of existing literature. Preliterate peoples are usually adept in the knowledge of the heavens: they have no other nightly reading matter. The regularity of eclipses is the subject of some of the earliest recorded scientific knowledge, perhaps as early as over 30,000 years ago, if Alexander Marshack's famous analysis of a palaeolithic tally stick is right.[37] Comets in some cultures are scrutinized as if they were starry messengers but there is no reason, apart from Harriot's assertion, to suppose that indigenous culture in sixteenth-century Virginia was among them. The notion that eclipses could be exploited to activate savages' superstitions was already a topos when Columbus claimed to have done so on Jamaica, when he was cast away there in 1504, in order to extort supplies from the natives. Eclipses and comets were among the omens that, in the 1540s, the Spanish Franciscan ethnographers, Tomás de Motolinía and Bernardino de Sahagún, adduced among the alleged reasons for the supposed collapse of Aztec morale in the face of Cortés's invasion. Both incidents owed, as far as we can tell, nothing to genuine indigenous beliefs, but were drawn from European sources – Herodotus, in Columbus's case, and works by Josephus, Lucan, and Plutarch in that of the Franciscans.[38] If Harriot was right about the demoralizing effects of such celestial phenomena among the potential victims of English colonization, it was thanks to an unverifiable assumption, shared with Spanish predecessors, that 'natives' are superstitiously prone to demoralization.

Harriot's assumption that Virginian natives mistook Englishmen for gods prompts similar reflections. '[S]ome people', Harriot averred, 'could not tell whether to think us gods or men ... Some therefore were of opinion that we were not borne of women, and therefore not mortall, but that we were men of an old generation many years past then risen again to immortality ...'[39] The similarity with well-known Spanish texts is again unmistakable. Columbus's claim that Tainos called him and his crew 'men from heaven',[40] which he made in despite of the absence of an interpreter and in default of any knowledge of indigenous language, became embedded in traditions about the receipt of Spaniards by Native American hosts, repeated with some variations in many successive accounts. Cortés abused the same

---

37 A. Marshack, *The roots of civilization* (1972), 95–108.
38 F. Fernández-Armesto, 'Aztec' auguries and memories of the conquest of Mexico', *Renaissance studies,* 6 (1992), 287–305.
39 Quinn (ed.), *Roanoke voyages*, vol. 1, 379–80.
40 C. Varela (ed.), *Cristóbal Colón. Textos y documentos completos* (Madrid, 1984), 104.

assumption by asserting that an Aztec paramount recognized him as a prophetically anticipated returning culture-hero,[41] whom writers of the next generation identified as Quetzalcoatl – the name of a deity (or perhaps more accurately of a cluster of divine attributes) – often depicted as the 'feathered serpent' of Mesoamerican pantheons.[42] Despite the inherent unlikelihood that my fellow-Spaniards could be mistaken for gods, and the lack of any supporting evidence, historical tradition has persisted, as it persisted in Harriot's day, in ascribing conquistador success to indigenous superstition in this respect.[43] Whether the Aztecs, in particular, had or even could possibly have had any such myth of a 'god from the sea' has been the subject of much scholarly dispute, as has the meaning of comparable myths among peoples of the Mexican highlands.[44] A similar debate echoes in literature about whether Captain Cook was similarly miscast by natives who put him to death in Hawai'i.[45] For reasons we shall come to in a moment, I think it likely that Spaniards genuinely thought that some natives divinized them, especially in coastal cultures, where analogous phenomena have been said to occur with sufficient frequency to command acknowledgement, or where inhabitants associated strangers from afar with what the anthropologist Mary W. Helms has called 'the touch of the divine horizon'.[46] The notion, however, that in any such places resistance faltered as if in submission to gods has no evidence to support it. Again, conquistador literature seems to have lulled Harriot into excessively sanguine assumptions and expectations.

It may fairly be asked why Spanish traditions included so many traps for unwary Englishmen. If Spanish prowess, European technology, and native superstition played little or no part in the founding of the Spanish empire, why did early narrators give them so much prominence, and, if not by such means as I have dismissed or marginalized, how did the empire come into being? The first of these questions has already been answered in part: prowess was part of an image conquistadores projected of themselves, partly to claim rewards from the crown and partly to reflect the literary traditions they acquired from reading chivalric romance. The story of the origins of the Spanish empire was not, however, written only by the victors. It may also have suited natives' susceptibilities in early colonial times to nourish the myth that it took a race of supermen to subject their ancestors.

41 J.H. Elliott, 'The mental world of Hernán Cortés', *Transactions of the Royal Historical Society*, 17 (1967), 41–58.

42 J. Lafaye, *Quetzalcoatl and Guadeloupe. The formation of Mexican national consciousness* (Chicago, IL, 1976); M. Restall, *When Montezuma met Cortés* (New York, 2018), 43–45, 99–102.

43 Restall, *Seven myths*, 139-40.

44 H. Thomas, *Conquest* (New York, 2005), 178-90; David Carrasco, *Quetzalcoatl and the ironies of empire* (Boulder, CO, 2000).

45 M. Sahlins, *How 'natives' think. About Captain Cook, for example* (Chicago, IL, 1995).

46 M. W. Helms, *Ulysses' sail* (Princeton, NJ, 1988); *Craft and the kingly ideal* (Austin, TX, 1993).

Clerical writers, like Motolinía and Sahagún, contributed at least as much to the early historiography as the conquistadores themselves and the secular writers whom they or the lay Spanish administrators encouraged, patronized, and informed. Indigenous authors proposed versions of their own in the service of their own agendas. Authors of sources of all three types of provenance shared one objective, which I think more recent historians have failed to appreciate: in early colonial times everyone found the outcome of the conquest baffling. Indigenous communities were often uncomprehending and resentful of the fact that the fall of bygone hegemonies, such as those of Aztecs and Incas, should have benefited outsiders. Spaniards, most of whom were escapees, of modest social backgrounds, from a world of restricted opportunity at home, were astonished at the change of fortune that had elevated them to supremacy in a world in which they were vastly outnumbered and ill equipped by culture, language, and education to understand. In their recourse to myths of heroic prowess, returning gods, dispiriting omens, and technical wizardry we can read the extent of their surprise. Clerical authorities, however, had a distinct agenda, which some indigenous writers shared. It did not suit them to credit the conquistadores with a victory owed to human heroism. The clergy were competitors with the conquistadores for control of indigenous lives and labour. They therefore proposed an alternative myth: that Providence, manifest in miracles, procured the favourable outcome Spaniards enjoyed.[47] Natives found the explanation appealing, because on the whole it is more creditable to be defeated by God than man. Conquistadores subscribed to the notion that miracles, especially in the form of battlefield apparitions, hallowed and legitimized their depredations. Harriot may seem an unlikely person to be suckered by the supernatural. Enemies notoriously accused him of atheism, and the exact nature of his understanding of God, as previously lecturers in the present series have pointed out, is hard to pin down.[48] But Harriot had no difficulty in appropriating Spanish tradition in this connection, switching it to dignify his own fellow countrymen with divine favour, ascribing to some of his comrades the claim to providential intervention, and professing his assent. '[S]ome', he stated of the early success the English invaders enjoyed, 'said that it was the special work of God for our sakes, as we ourselves have cause in some sort to thinke no less'.[49]

Harriot also coincided with Spanish predecessors in advancing claims that Christianity would confirm a new order of alien supremacy. Natives' permanent

---

47 Restall, *Seven myths*, 133-35.
48 S. Clucas, 'Thomas Harriot and the field of knowledge in the Renaissance', in Fox (ed.), *Thomas Harriot. An Elizabethan man of science*, 93–136; S. Mandelbrote, 'The religion of Thomas Harriot', ibid., 246-79. See also J. Jaquot, 'Thomas Harriot's reputation for impiety', *Notes and records of the Royal Society of London*, 2 (1952), 164-67.
49 Quinn (ed.), *Roanoke voyages*, vol. 1, 380.

submission, he suggested, would be a consequence of 'the imbracing of the trueth'. He also echoed Spanish texts in expressing confidence that natives had the advantage of a sort of partial pre-revelation from God to aid and ease their reception of the new religion. '[T]hey may', he suggested, 'in short time be brought to civility, and the embracing of true religion', partly because '[s]ome religion they have already, which although it be far from the truth, yet being as it is, there is hope it may be the easier and sooner reformed. They believe that there are many Gods ... but ... one only chief and great God'.[50] The search for evidence of a pre-revelation or partial revelation of the nature of God was a longstanding thread in Spanish missionaries' work in the New World. To demonstrate that in-digenes could easily be converted was a crucial missionary strategy in com-petition with lay Spaniards for access to native people, since to be 'on the way to conversion' was one of the conditions that exempted pagans from en-slavement in a decretalist tradition, explicit in a Bull of Eugenius IV.[51] Among claims missionaries made were that natives had practices that pre-figured sacraments, including confession and marriage; that they stoned adulterers in conformity with the customs of Jews in the time of Christ; that the training of their priests resembled Franciscan methods,[52] and, as in Harriot's text, that they had notions corresponding to that of a supreme deity – ne-cessarily knowable by nature, according to Las Casas.[53] Identifications of Quetzalcoatl with St Thomas and allegations of evidence that St Thomas had visited America in antiquity served a similar purpose of showing that American ground was laid for missionary work.[54] Though Harriot's ob-jectives had little in common with Spanish priests', it made sense for him to echo their perceptions or claims in this respect. Christianization was a path both to pacifying the natives of Virginia and to securing their consent to English rule.

Harriot's account of his own foray into missionary work includes a cur-ious echo of or resemblance to traditions about a famous episode in the history of the conquest of Peru. When Pizarro seized Atahualpa and mas-sacred his attendants at Cajamarca in 1532, the slaughter was triggered by the Inca's dismissal of the Bible, apparently because, unfamiliar with European methods of recording information, he did not realize that it was other than a valueless material object, unworthy to be accepted as a gift.

---

50 Ibid., vol. 1, 372.
51 Muldoon, *Popes, lawyers and infidels*, 121-24; 'Crusading and canon law', in H. Nicholson (ed.), *Palgrave advances in the Crusades* (Basingstoke, 1975), 37–57.
52 D. de Landa, *The Maya. Diego de Landa's account of the affairs of Yucatán*, ed. A. Pagden (Chicago, IL, 1975), 74–75, 79–80; F. Berdan and P. Anawalt (eds.), *The essential Codex Mendoza* (Berkeley, CA, 1997), 172–83.
53 Bartolomé de Las Casas, *In defense of the Indians*, ed. S. Poole (DeKalb, IL, 1974), 239–40.
54 Lafaye, *Quetzalcoatl and Guadeloupe*; David Brading, *The first America* (Cambridge, 1991), 255–92.

According to some accounts, he examined it perfunctorily, in an apparent attempt to 'listen' to what it supposedly had to say, and flung it contemptuously aside.[55] The relevant passage in Harriot's text says:

> I made declaration of the contents of the Bible; that therein was set forth the true and only GOD, and his mighty works, that therein was contained the true doctrine of salvation through Christ, with many particularities of Miracles and chiefe points of religion, as I was able then to utter, and thought fit for the time. And although I told them the booke materially & of itself was not of any such virtue, as I thought they did conceive, but only the doctrine therein contained; yet would many be glad to touch it, to embrace it, to kiss it, to hold it to their breasts and heads, and stroke over all their body with it; to show their hungry desire of that knowledge which was spoken of ...[56]

The last respect in which Harriot's Virginia resembled the New World described in Spanish accounts of conquest was in the outbreak of disease. Harriot anticipated many modern historians in supposing that the effect, subversive of native morale, was to make conquest easier:

> The disease also so strange, that they neither knew what it was, nor how to cure it; the like by report of the oldest men in the country never happened before, time out of mind. A thing specially observed by us as also by the naturall inhabitants themselves. ... Insomuch that when some of the inhabitants which were our friends ... had observed such effects in four or five towns to follow their wicked practices, they were persuaded that it was the work of our God through our meanes, and that we by him might kill and slay whom we would without weapons and not come near them.[57]

Harriot did not have to rely on Spanish texts for such observations. The results of native exposure to unfamiliar pathogens was no doubt similar in Virginia to those in most regions of the New World. Whether Harriot and historians who have made similar inferences were right in supposing that disease was, in effect, the invaders' ally is more open to question. The inference must be judged alongside other considerations. Spaniards themselves, for instance, were susceptible to features of a new disease environment and to the debilitating conditions of previously unexperienced environments and diets.[58] Since they relied heavily on native allies or auxiliaries, who bore the

55 Sabine MacCormack, 'Atahualpa and the book', *Dispositio,* 14 (1989), 141–68.
56 Quinn (ed.), *Roanoke voyages,* vol. 1, 376–77.
57 Ibid., 378-79.
58 H. Figueroa Marroquín, *Enfermedades de los conquistadores* (Guatemala City, 1983).

brunt of the fighting in most encounters, and who were far more exposed to diseases endemic among the newcomers, the impact of disease might be expected to have been more adverse than advantageous to them. The lethality of 'the breath of a Spaniard' might inspire desperation of more than one kind, stirring resistance rather than resignation. In the case of Iroquoian resistance to French intruders in the seventeenth and eighteenth centuries, awareness of the enemy as vectors of deadly microbes seems to have made natives anxious to exterminate or expel them as quickly and thoroughly as possible.[59] In any case, in practice Harriot seems to have been wrong. Virginian natives, though often disposed to be welcoming, could be driven to resist, and disease does not seem to have deterred them.

Harriot's expectations, or professed expectations, of an empire easily founded did not arise only, of course, from his interpretations of Spanish experiences and readings of Spanish texts. In some ways, experiences the English intruders recorded conduced to the same end: the 'kind and loving reception' they first enjoyed; the success of their first attempts to trade, when they established dazzlingly favourable exchange rates, giving a modest kettle for skins worth fifty crowns.[60]

Harriot's illusions are pardonable, as historians have shared them, and even the Spaniards who took part in the conquest and inhabited early colonial society failed to comprehend the reasons for their own success. It is in retrospect hard to understand how so many subject communities, whose grasp was, at best, imperfect of who and where the king of Spain was, and of what his claims to legitimacy and authority were worth, should have aligned themselves with Spaniards, who rarely seemed to show much comprehension of native values and priorities and often engaged in provocations likely to breach the peace. To approach understanding of how to found an empire in Harriot's day, we have first to acknowledge the point with which the present piece started: that empires – especially pre-industrial empires – are weak; broadly speaking, their power diminishes with distance, so that their authority is feeble at remote margins; and that in consequence metropolitan rulers or elites depend on regional and local collaborators. Nor can we hope to understand what happened if we cling to images of irrational or inferior beings submitting to superior white men. Unless we acknowledge that native peoples retained their initiative in the face of would-be conquerors and made rational decisions, consistent with their own culture and traditions, we shall never grasp the origins of the modern Americas.[61] Generally speaking,

---

59 D.K. Richter, *The ordeal of the Longhouse* (Chapel Hill, NC, 2011), p. 50; Karl H. Schlesier, 'Epidemics and Indian middlemen: rethinking the wars of the Iroquois, 1609–1653', *Ethnohistory*, 23 (1976), 129-45.

60 Quinn (ed.), *Roanoke Voyages*, vol. 1, 101.

61 Gonzalo Lamana, *Domination without dominance. Inca-Spanish encounters in early colonial Peru* (Durham, NC, 2008); Nancy Farriss, *Maya society under colonial rule* (Princeton, NJ,

Spaniards founded their empire where they were favourably received. Where they were not, it commonly took generations of fighting to subdue resistance or, in many places, reach a stalemate or admit defeat. In some places, the difference was a matter of calculation on the part of natives who saw advantages in adhesion to empire. All indigenous peoples had traditional hatreds and conflicts of their own with neighbours, as Harriot noted in the Virginian case.[62] They made use of newcomers for their own purposes in the course of their internecine wars.

For other reasons, in many cases they deferred to the intruders and placed them in positions of command in war and authority afterwards. Some cultures, such as were common in the Americas, have a propensity to receive strangers with exceptional honour. I call this phenomenon the stranger-effect.[63] In modern Western societies the propensity is rare. We mistrust strangers. We reject them. In the USA some of us call them illegals. We impose on them bureaucratic or fiscal burdens. If we admit them, we often make them unwelcome and typically assign them low status and demanding or demeaning work. In other times, however, and in other parts of the world, people have behaved differently. Sacred rules of hospitality oblige people in some cultures to greet strangers with their best gifts and goods and women and even actual deference. When Spaniards found themselves treated in this way in parts of the Americas it made them feel godlike: hence, I suggest, the myths of the return of Quetzalcoatl *et hoc genus omne*. A possibly useful analogy is with the value added to exotic goods and long-range imports in modern Western commerce. So it is, in many cultures, with people from afar. In Christendom in the past, pilgrims profited from a similar effect, acquiring prestige with neighbours on returning home.

To defer to the stranger – given an appropriate cultural context – is often a highly commendable, rationally defensible approach. The stranger is useful as an arbiter or judge because he or she is uninvolved in existing factional and dynastic conflicts and can bring an objective eye to matters in dispute. The early colonial archives of Spanish America are full of cases in which native elites confided in Spanish arbitration, gradually thereby shifting power into Spanish hands.[64] For similar reasons, strangers, untainted by prior associations with local rivalries, make first-class allies, bodyguards, and close counsellors for rulers and marriage partners for elite families. A case in point, which vividly illustrates the operation of the stranger-effect, is that of Alonso de Illescas and his companions, castaways in a shipwreck on the coast of

1984); J.E. Kicza, *Resilient cultures. America's native peoples confront European colonization* (Upper Saddle River, 2003).

62 Quinn (ed.), *Roanoke voyages*, vol. 1, 371.
63 F. Fernández-Armesto, 'The stranger-effect in early modern Asia', *Itinerario*, 24 (2000), 80–103.
64 M. Restall, L. Sousa, and K. Terraciano (eds.), *Mesoamerican voices* (Cambridge, 2005), 47–54, 64–96, 101–13.

Ecuador in 1553. The newcomers followed the typical trajectory of Spanish conquistadores: welcome among locals who fed them, employed them as allies in war, confided to them the care of their paramount, espoused their leaders as the consorts of princesses, and eventually handed them supreme power. In this case, however, the intruders had none of the advantages of superior weapons or learning or divine resemblance, or providential favour or thaumaturgic power that chroniclers and historians have ascribed to Spaniards or Harriot hoped for in help of an English empire. For Alonso and his fellow-adventurers were black slaves, who had nothing to speed them towards supremacy except the goodwill of their hosts.[65]

In one respect, Harriot interpreted accurately the lessons of Spanish experience. Like Hakluyt and Ralegh, he was aware of the Black Legend of Spanish cruelty and anxious not to incur an English equivalent. But the legend, though not entirely false, was itself delusive, suggesting that the balance of power in colonial society was all on the intruders' side and that their obligation to show mercy and clemency arose from a position of supremacy. The reverse was the case. In their empire, Spaniards' weakness, if tested by a show of native unity in resistance or revolt, usually cracked, as it did when Pueblo rebels drove Spain from New Mexico in the 1680s, or stalled, as it did until the late eighteenth century on the Mapuche and Comanche frontiers. When Spaniards triumphed, they always had native collaborators on their side. English dependence on native goodwill was obvious when a collapse of good relations, which Harriot deplored, occurred towards the end of his time in Virginia, when English demands for supplies, willingly proffered at first, became intolerable and native allies refused them. 'Some of our company', he lamented, 'towards the end of the year showed themselves too fierce, in slaying some of the people, in some towns, upon causes that on our part might easily enough have been borne withal'.[66] Harriot appreciated that 'discreet dealing and government' was essential to the peaceful establishment of an English colony, but he assumed that such discretion would be exercised from a position of strength. He was, by the standards of English invaders, and perhaps of lay Europeans generally, exceptionally sympathetic to the natives he met, learning to converse in their language, expressing appreciation of their 'wit' and ingenuity.[67] Yet he regarded the outbreak of violence as provoked by the natives' unreasonableness. He expected them to maintain subservience, not realizing, I think, that the stranger-effect, not English superiority, was the basis of co-operation

65 C. Beatty-Medina, 'Alonso de Illescas', in Beatriz Gallotti Mamigonian and Karen Racine (eds.), *The human tradition in the black Atlantic, 1500–2000* (Lanham, 2010), 9–23; 'Caught between rivals: the Spanish-African Maroon Competition for Captive Indian Labor in the Region of Esmeraldas during the late sixteenth and early seventeenth centuries', *The Americas,* 62 (2006), 113–36.
66 Quinn (ed.), *Roanoake voyages*, vol. 1, 381.
67 Ibid., vol. 1, 371.

between newcomers and indigenes, and that it was a wasting asset. In the *Brief Report* he professed to retain 'good hope' that despite the excesses of outrage that had imperilled his expedition, natives would come 'to honour, obey, feare and love us'. The next century of English endeavours in Virginia showed that his successors had what would nowadays be called a long 'learning curve' ahead of them. Fear and love figure, for sure, among emotions exchanged between communities in colonial societies; but the only means of founding an early modern empire were perceived community of interest between incumbent and intrusive elites. The stranger-effect was the surest way of initiating friendly relations.

7 Thomas Harriot's Magnificent Book: Creating Europe's First Illustrated Exploration Narrative

Larry E. Tise

On 15 May 1820, the Clerk Assistant at Table of England's House of Commons, John Rickman (1771–1840), penned a brief letter to his friend, the Revd Thomas Sockett (1777–1859), rector of the church at Petworth in Sussex. He had a simple request. A renowned statistician and master of precise information, Rickman sought Sockett's assistance in finding a hand-colored map drawn by Thomas Harriot, believed now to be in the possession of George Wyndham (1751–1837), third Earl of Egremont. Prompted by a gentleman he described as 'the Agent of the N[orth] Carolina government', Rickman explained, 'he wishes for a Copy of the Coloured Print' because it contains a map 'of the precise spot about which his government have reason for anxious research'. Persuaded the North Carolinian's request was credible and worthy, the busy Rickman, who channeled requests to Members of Parliament daily, suggested 'perhaps Lord Egremont would upon your Request [Rev. Sockett's] extend his kindness so far as to permit a Copy to be taken'. To advance the process, Rickman also offered the help of an expert copyist: 'I shall endeavor to prevail on Miss Marianne Daysh[?] to perform this Service if permitted'.[1]

The agent of the State of North Carolina who prompted this 'anxious research' was an esteemed attorney, Peter Browne (1766–1833), leader of a campaign to build canals and navigation improvements in the same territory explored and settled by Sir Walter Ralegh in the sixteenth century. A native of Knockadock in Aberdeenshire, Scotland, Browne arrived in

---

1 Letter, J[ohn] Rickman, Palace Yard [Westminster] to Rev. T[homas] Sockett, Petworth, 15 May 1820, inserted in William Strachey manuscript 'The first Booke of the first Decade contyning the histories of travels into Virginia Britania, expressing the Cosmographies, & Commodities of the Countries, together with thos Qualities, Customes, and Manners of the naturall Inhabitants', addressed 'To the Right honorable and accomplisht great Lord covetous of all Knowledg Henrye Earle of Northumberand, Esq.' (ca. 1612), Special Collections, Firestone Library, Princeton University. Both Rickman and Sockett were well-known figures in the 1820s. For Rickman, see https://en.wikipedia.org/wiki/John_Rickman consulted 7 August 2020. For Sockett, see S. Haines and L. Lawson, *Poor cottages & proud palaces. The life and work of the Reverend Thomas Sockett of Petworth, 1777–1859* (Hastings, 2007).

DOI: 10.4324/9781003096580-8

North Carolina as a young man, friendless and destitute. But he somehow rose from obscurity to prominence in both law and business in the state's capital city, appropriately named Raleigh to honor the Elizabethan empire builder. At age forty Browne personally financed publication of the first detailed map of the State of North Carolina. At age fifty he chaired the Internal Improvements Commission of North Carolina. At fifty-two he decided to take his ample wealth and retire to his native Scotland.

But filled with curiosity about the disappearance of those navigable inlets used by Ralegh's explorers in the 1580s to reach Roanoke Island, Browne began his search in Oxford where he was put on the trail of Lord Egremont. Two weeks before contacting Rickman at Westminster Palace, he wrote his chief collaborator in building North Carolina canals: 'Both the University of Oxford and Earl Egremont have evinced the utmost readiness to serve us with regard to Mr. Harriot's survey'. Though in hot pursuit of the original map, he cautioned: 'I fear it is not now in existence; this however is not yet ascertained to a certainty'. Yet Browne, convinced that he would find the grail, ended his letter 'I think I shall be with you before snow', thus ending his brief Scottish retirement. He took his discoveries back to North Carolina, where he spent the rest of his life promoting canals and navigable waterways – including an effort to restore the inlet shown on Harriot's hand-colored map of the Carolina coast (see Figure 1).[2]

Peter Browne's quest to find this hand-colored map of North Carolina's inlets in the 1580s is but a single vignette of the thousands of researchers, designers, publishers, and curiosity-seekers who have sought out the beautifully engraved 1590 edition of the *Briefe and true report of the new found land of Virginia*. Alongside the great atlases of Abraham Ortelius, the illustrated edition of Harriot's book became the most influential book of the era of European exploration. It was among the most frequently printed and distributed books of the age. It shaped the way we have envisioned the bodies and habits of American Indians from the sixteenth century to the present. The book inaugurated the use of copper plate engravings to present full body portraits and action scenes of living indigenous persons and their communities. It introduced a narrative text (written by Harriot) and engravings based on original art produced on site revealing the nuances of real faces, postures, and indigenous dress. It was, in fact, the inaugural volume in a folio-sized series of illustrated books known as 'Grands Voyages' designed and published in Frankfurt, Germany, the book capital of Renaissance Europe.

2   Letter, P[eter] Browne, London, to Archibald D. Murphey, Haw-River, North Carolina, 27 April 1820, in W. H. Hoyt (ed.), *The papers of Archibald D. Murphey*, 2 vols. (Raleigh, NC, 1914), vol. 2, 163–64; also biographical entry on Peter Browne, ibid., vol. 1, 80–81, n. 2. For further biographical information on Browne, see https://www.ncpedia.org/biography/browne-peter, accessed 7 August 2020.

Figure 7.1 Plate II. "The arrival of the Englishmen in Virginia" Thomas Harriot, *A briefe and true report of new found land of Virginia* (1590). This is the hand-colored map (shown here in black and white) of the Carolina barrier islands indicating inlets and shallow waters in 1590 sought by Peter Browne in the collections of George Wyndham, 3rd Earl of Egremont at Petworth House in May 1820. The map is contained in the William Strachey manuscript "The first decade conteyning the historie of travel into Virginia Britania" now located in the Princeton University Firestone Library's Rare Books & Special Collections (C0199). Courtesy of Firestone Library Rare Books & Special Collections.

In addition to being a breath-taking departure in producing books, the 1590 edition of the *Briefe and true report* catapulted Thomas Harriot into a position as one of the most important authors of the sixteenth and early seventeenth centuries. Thomas Harriot a renowned author? Yes, an internationally re-cognized author who invented a methodology and format for analyzing and reporting anthropological wisdom and exactitude, even objectivity, in an essay on the American Indians he encountered on the Carolina coast. He wrote his original text as a popular account of what he saw in America – without pictures. But he wrote his report with the precision of a social sci-entist, making it superbly suited for illustration. The initial printing of his tract was in the form of an inexpensive quarto designed to refute rumors that the Ralegh expedition of 1585–86 was a failure. But with Harriot's skill and de-termination, it was reissued less than two years later as a lavishly illustrated folio on quality paper and simultaneously printed in four languages (Latin, German, French, and English). A great accomplishment for an author in any age and at any age (he was 30 years old), especially in the earliest days of copper plate engraving.

The landmark achievement of four simultaneous versions of the 1590 edition of *Briefe and true report* contrasts sharply with an often-heard regret among Harriot scholars that he failed to publish anything worthy of note during his lifetime. Also, observations that Harriot either lacked the drive or a suitable patron to bring any of his great researches to print. These sighs of woe, repeated so frequently among Harriot essayists and biographers, have unfortunately, if unintentionally, projected an image of the man as something of a failed being for whom we must apologize before we begin describing his intellectual achievements. One of Harriot's greatest cheerleaders, historian John W. Shirley (1908–88), opened his superb *Thomas Harriot: A biography* (1983) with a 38-page chapter on why his subject had been overlooked for three and a half centuries. After devoting his career to finding every document relating to Harriot, Shirley barely mentioned his greatest achievement.[3]

Sadness about Harriot's publication shortcomings have also been heard at the esteemed Thomas Harriot Lectures at Oriel College, Oxford. Sorrowful elegies can be read in abundance in the first two volumes of these lectures, edited by Robert Fox: *Thomas Harriot: an Elizabethan man of science* (2000) and *Thomas Harriot and his world* (2012). This was also the case in an earlier international conference organized by Shirley in 1971 at the University of Delaware and published with Shirley as editor as *Thomas Harriot: Renaissance scientist* (1974).[4] The most frequently referenced causes of Harriot's failure to publish can be categorized as: (1) His evident disinterest in publishing his own research; (2) His absence of a patron during the years he was conducting scientific inquiries (i.e., Ralegh and the Earl of Northumberland were in the Tower of London); (3) He worked mainly as a loner and may not have known that he had made unique discoveries; and (4) those individuals to whom he left his papers did not carry out his wishes for his work to be published.[5]

Harriot biographer Robyn Arianrhod makes a similar assessment of the *Briefe and true report* in her 2019 biography of Harriot. She follows a long habit of bypassing the first printing of the tract as 'his 1588 pamphlet'. But

3 J. W. Shirley, *Thomas Harriot. A biography* (Oxford, 1983), 1–37. Shirley also lamented that Henry Stevens, the renowned book-dealer who sold more Harriot books than anyone ever, failed to publish his own biography of Harriot during his lifetime. But he praised Stevens for noting that the 1588 pamphlet version of *Briefe and true report* was 'one of the outstanding rare books of all time' (ibid., 27–30).

4 R. Fox (ed.), *Thomas Harriot. An Elizabethan man of science* (Aldershot and Burlington, VT, 2000). R. Fox (ed.), *Thomas Harriot and his world. Mathematics, exploration, and natural philosophy in Early Modern England* (Farnham and Burlington, VT, 2012). J. W. Shirley (ed.), *Thomas Harriot. Renaissance scientist* (Oxford, 1974).

5 The author has assembled a brief compilation of the chief lamentations contained in the lectures appearing in these volumes.

she nevertheless notes Harriot'd authorship of the captions to White's drawings in the 1590 version which was 'published in deluxe editions in four languages: Latin, English, French, and German' with 'engravings of Whites drawings and paintings made by Theodor de Bry'. Further that the book 'was an international hit ... making Harriot, White, and Ralegh famous in Europe'. Good praise for some instant authorial fame. But these bare-bone notations understate the revolutionary character of Harriot's achievement and its enduring impact upon the way we have imagined American Indians for over four centuries.[6]

By 1662, 31 years after Thomas Harriot's death, it had become clear that Nathaniel Torporley, an executor of his last will and testament, had failed to publish the mathematical and scientific papers of the great mathematician. Harriot designated Torporley both as executor and 'Overseer of my Mathematicall Writings ... to peruse and order and to separate the Cheife of them from my waste paper, to the end that after hee doth understand them he may make use in penning such doctrine that belonges unto them for publique uses'. Concerned that Harriot's scientific and mathematical discoveries might be lost forever, the Royal Society began a search for the whereabouts of the papers transferred to Torporley. Coming only two years after the prestigious Society came into existence, the search which persisted for years underscored the honored stature of Harriot in the historic pantheon of the kingdom's great mathematicians. But the Society was unable to resolve the issue of what papers were in the trunk turned over to Torporley, how they may have been organized, and what discoveries Harriot thought they would demonstrate when Torporley put them in print. And thus many quandaries about Harriot's theories and discoveries have remained to the present.[7]

In contrast to the contorted history of Harriot's scientific and mathematical papers, there has been little doubt about the history and continuous importance of his monumental *Briefe and true report*. From its first appearance in 1588 as a compact tract on Sir Walter Ralegh's ambitious exploratory colony of 1585–86, it took its place as a significant time capsule on the emergence of the English empire in North America. When it reappeared in 1590 as the first illustrated book of Theodor de Bry's 'Grands Voyages', its distinction as a cornerstone of European expansion literature magnified geometrically. This, too, has been known for centuries. But the crucial role played by Harriot in causing this to happen and the cascading impact of the book to the present are less well-known and documented. Although the distinguished historian David Quinn revealed much about Harriot's authorship, there is more to be documented, including an explosion of interest in the book in the twenty-first century.

6 R. Arianrhod, *Thomas Harriot. A life in science* (Oxford, 2019), 80, 89, 109.
7 Shirley, *Thomas Harriot. A biography,* 7–27, 460–70.

The crucial connecting links that put Harriot in a position to produce his book were his career-long employer and mentor Sir Walter Ralegh and his chief publicist Richard Hakluyt. The author of one of Ralegh's most extensive biographies describes Ralegh as the 'architect' of the English empire.[8] Hakluyt's principal biographer, meanwhile, describes his subject as having an 'obsession for an English America'.[9] Both Ralegh and Hakluyt found in Harriot the learned scientist, mathematician, cartographer, and onsite expert they needed to launch their empire. Despite a keenly active exploratory mind throughout his life, in the first phases of his career Harriot was a problem solver, a sharp observer, and a versatile applied scientist.

At the age of only 23 in 1583, Harriot was chosen by Ralegh to devise and teach courses in the science of navigation, making use of observational instruments and mathematical calculations to set courses of nautical travel and to determine latitude and longitude while at sea. These new skills deriving from Harriot's studies in arithmetic, geometry, and trigonometry at St Mary Hall were to supplement, not replace, the learned wisdom and experience of seasoned seafarers including widely available Portuguese pilots. How did the young Harriot come to Ralegh's attention? They both studied at Oxford, but years apart. Both, however, were probably known to Richard Hakluyt, whose years at Oxford overlapped those of Harriot. Hakluyt gave public lectures on geography at Oxford and was envisioning an English empire and justifications for it throughout his 16 years of study and generous pension support (1570–86) at the University. His publication of *Divers voyages touching the discoverie of America … made first of all by our Englishmen* (1582) brought him to the attention of the royal ministry surrounding Queen Elizabeth. Also to the attention of Ralegh, who was in the process of winning a patent for the exploration and colonization of North America. From 1583 forward Hakluyt, himself only 30 years old, acquired the mantle of theoretician and public voice for a burgeoning English empire. As editor, author, speaker, and organizer, he was directly or indirectly involved in every English colonial venture from 1583 until his death in 1616.[10]

With the blessings of both Hakluyt and Ralegh, Harriot is believed to have trained the prospective captains and pilots for all of Ralegh's ventures in the concepts and calculating skills for nautical navigation. That includes Ralegh's explorations of North America in the 1580s, his war-making exploits against the Spanish, and his first travels to Guiana in 1595. From the journal of the flagship *Tyger* on Ralegh's 1585–86 expedition to the Carolina coast, we know the names of many of the ships' masters likely trained by Harriot: 'Master Thomas Candishe, Master John Arundell, Master Raimund, Master Stukely,

---

8 A. Gallay, *Walter Ralegh. Architect of empire* (New York, 2019), e.g., 11–25.

9 P. C. Mancall, *Hakluyt's promise. An Elizabethan's obsession for an English America* (New Haven, CT, 2007), e.g., 128–55.

10 Mancall, *Hakluyt's promise*, 25–101.

Master Bremige, Master Vincent, and Master John Clarke, and divers others, whereof some were Captains, and other some Assistants for counsel, and good directions in the voyage'. Philip Amadas and Arthur Barlowe, veterans of the prior 1584 expedition, were also among the large company. Our historical documents provide little information on Harriot's involvement with navigational decisions on the 1585–86 transoceanic passages or across the inlets and sounds of Carolina's barrier islands. But in this single seagoing expedition of his life, he was surely near the pilot's cabin, testing his navigational theories and the accuracy of his nautical instruments.[11]

We also know that Harriot made extensive scientific preparations for what became a full year of observation and analysis of everything he and his associates encountered on the Carolina coast, its sand barriers, and its largely swampy, submerged terrain. He brought a set of instruments sufficient to produce an astonishingly accurate map of the barrier islands and the many inlets that dotted that complicated environment, plus a supply of paper and writing materials to record observations and measurements in the field. He charted accurately the principal waterways that fed into the Carolina sounds and the locations of Indian settlements. Among the items in his drafting chest were 'a universal dial', 'a cross staff', 'a sailing compass', instruments 'for the variation of the compass' and 'declination of the needle', *Ephemerides* or 'other calculated tables', 'paper royal', quills, ink, 'black powder to make ink', a 'stone to grind colours', and 'all sortes of colours to draw all things to life', 'two pairs of "brazen Compasses"', and 'other Instrumentes to drawe cardes and plottes'.[12]

One of the principal goals of the 1585–86 expedition was to identify plants to produce dyes and medicines, trees for cordage in building ships, and abundant food plants. Harriot took along the finest illustrated guide to flora, food plants, mosses, and trees the English language. That was the newly translated bible of plant life during the era, Nicolas Monardes, *Joyfull newes out of the newe founde worlde ... Englished by John Frampton* (1577). With this surprisingly accurate glossary of plants Harriot was equipped to compare the plants he encountered with similar species in both Europe and the Indies. For example, when he encountered the plant sassafras, he was able to give the local Indian name for it – *winauk* – and 'for the description, the maner of using, and the manifold vertues thereof, I refer you to the booked Monardes, translated and entituled in English, *The joyfull newes from the West Indies*'. When he came to describing corn (or *pagatowr* in Algonquin) he turned again to Monardes: 'A kinde of graine ... the same in

11 'The Holinshed account of the 1585 expedition', in David Quinn (ed.), *The Roanoke voyages, 1584–1590*, 2 vols., continuously paginated (London, 1955), vol. 1, 173-75; 'The Tiger journal of the 1585 voyage', ibid., vol. 1, 179-80. For his responsibilities in teaching and assisting with navigation, see ibid., vol. 1, 36–37, 52–53, 119; vol. 2, 513–15.

12 Quinn (ed.), *Roanoke voyages*, vol. 1, 52–53.

the West Indies is called Mayze; English men call it Guinny wheate or Turkie wheate, according to the names of the countreys from whence the like hath been brought'. He used the same book to identify what the Indians called *ascopo*. According to Harriot 'the barke is hoat in taste and spicie ... very like to that tree which Monardus describeth to be Cassia Lignea of the West Indies' – a kind of sweet bay tree used for the treatment of syphilis.[13]

Harriot took a similar premeditated approach to finding rare and useful minerals in the geological firmament of the Carolina coast. Among the personnel recommended for this expedition was 'an alcamist ... to trye the mettaylls that may be discovered'. A highly experienced metallurgist, Joachim Ganz (listed as Doughan Gannes in the personnel roster), joined the expedition and was called upon regularly to assay minerals and ores to determine their identity and quality. From a Jewish family in Prague, Ganz's father was an astronomer and historian, inspiring his son to a life of scientific inquiry. Carolina Indians wore copper jewelry, bracelets, and plates of copper suspended from their necks as a sign of rank. Ganz confirmed the quality of the Indians' copper supporting their boast of melting and retooling the metal over a simple wood fire. Ganz also confirmed the quality of iron ore found among stone rubble in the back country. While many early English adventurers were led astray by wrongly identifying minerals, that was not to be with the presence of Ganz.[14]

Perhaps the most important preparation for Harriot's scientific journey to North America was his deliberate effort to acquire the language skills he would need in order to communicate with the Algonquian Indians who populated the Carolina coast. Ralegh's 1584 explorers, Philip Amadas and Arthur Barlowe, returned to England in late September that year bringing two Algonquian Indians described by one observer as 'in countenance and stature like white Moors'. They were promptly paraded at Queen Elizabeth's court and elsewhere as curiosities from 'an island which is said to be larger than England'. Their 'usual habit [dress] was a mantle of rudely tanned skins of wild animals, no shirts and a pelt before their privy parts'. But to meet with Queen Elizabeth they were 'clad in brown taffeta'. The same observer commented: 'No one was able to understand them, and they made a most childish and silly figure'.[15]

While some Englishmen were gawking at them, Harriot set up living quarters for the Indians, Manteo and Wanchese, in the grand Durham House on the Thames where he lodged with Ralegh and the rest of the braintrust assembled to build an English empire. From late September 1584 until early April 1585 when he left with Sir Richard Grenville, heading to

---

13　Thomas Harriot, *Briefe and true report on the new found land of Virginia*, in Quinn (ed.), *Roanoke voyages*, vol. 1, 329 & n. 5; 338 & n. 1; 341 & n. 1; 345 & n. 1; 347 & n. 1; 366 & n. 1.

14　Quinn (ed.), *Roanoke voyages*, vol. 1, 136–37, 196, 274, 331–32.

15　Victor von Klarwill (ed.), *Queen Elizabeth and some foreigners* (London, 1928), 323.

America with Manteo and Wanchese in tow, Harriot learned to converse with them in both Algonquian and English. The bilingual conversations continued another three months as Grenville's flotilla made its way through the West Indies to the Carolina coast. Although Wanchese broke rank with the English expeditionary force when the convoy reached mainland North America, a regular interchange between Harriot and Manteo continued throughout the yearlong expedition. Since Manteo returned to England with Harriot in July 1586, the two men were close at hand for another ten months before his Indian companion returned to Roanoke Island with Ralegh's John White colony in May 1587. While historians often treat Harriot's interest in the Algonquian language as an exotic curiosity, Harriot and Manteo lived in almost immediate proximity for more than two and a half years. While English became a second language for Manteo, Algonquian was a well-practiced second language for Harriot as well.[16]

With his growing knowledge of Algonquian, Harriot became the most critical member of the 1585–86 expedition – the principal information-gathering foray of Ralegh's American ventures. He was the only member of the exploration party who could converse with Indian chiefs and warriors to secure vital information about geographic features and directions; Indian names for places, plants, and animals; and Indian reactions to English goods, maneuvers, and tactical blunders. But also intangible Indian perceptions about nature, life and death, and the gods. All of this was crucial to his role as chief scientist, principal investigator, and diplomatic advisor. As chief scientist, he determined the observational and experimental agenda during an entire year. As principal investigator, he assured that data were collected and recorded on plants, animals, fish, herbs, medicinals, diseases, people, weaponry, watercraft, and on such personal human characteristics as body marks, tattoos, jewelry, habits, and customs. His role as diplomatic advisor was perhaps the most demanding of all, since the expedition commanders were mainly crusty veterans of English maneuvers to suppress Irish rebels.

We know least of all about Harriot's role in the third category as diplomatic advisor. He is barely mentioned in the two surviving accounts of the 1585–86 expedition. In the first of these (*Tyger* journal), he accompanied 'Generall Grenville' on a grand tour of three Indian villages, one of which Grenville ordered to be burned over the disappearance of a silver cup. In the

---

16 Key dates in this timeline are as follows: Manteo and Wanchese arrived in England with Amadas and Barlow in mid-September 1584; Grenville's voyage departed Plymouth on 19 April 1585; the Drake rescue fleet left the Carolina coast on 27 July 1586; and the John White colony departed Plymouth on 8 May 1587. For Manteo and Wanchese's residence at Durham House, see Quinn (ed.), *Roanoke voyages*, vol. 1, 119; 'Thomas Harriot and the new world', in Shirley (ed.), *Thomas Harriot. Renaissance scientist*, 38–41; and Arianrhod, *Thomas Harriot,* 58–61. For a glossary of Algonquian words deriving primarily mainly from Harriot's use of these words, see J. A. Geary, 'The language of the Carolina Algonkian tribes', in Quinn (ed.), *Roanoke voyages*, vol. 2, 884–900.

second (Ralph Lane's discourse), Harriot was presumed to be present when Lane and his troops assassinated the powerful Indian chief Pemisipan (formerly Wingina) by beheading. Whether Harriot cautioned Grenville or Lane against these punitive acts we do not know. But Harriot revealed his view of these ill-advised military abuses in his published report, writing that 'some of our companie towards the ende of the year, shewed themselves too fierce, in slaying some of the people, in some towns, upon causes that on our part, might easily enough have bene borne withal'. He also advised future English colonies that the Indians 'in respect of troubling our inhabiting and planting, are not to be feared ... but that they shall have cause both to feare and love us, that shall inhabite with them'. As diplomatically as he could put it, his English associates were more likely to initiate bloody conflict than were the Indian inhabitants.[17]

Despite his need to use diplomacy on the principal failing of the exploratory venture, Harriot waxed with eloquence and authority on his scientific analysis of the flora and fauna and the natural resources of the Carolina coast and in an insightful anthropological assessment of its inhabitants. He began his treatise acknowledging he 'thought it good, being one that have beene in the discoverie, and in dealing with the natural inhabitants specially imploied; and having therefore seen and knowne more then the ordinarie; to impart so much unto you of the fruites of our labours'. Harriot divided his 'treatise' into 'three special parts'. The first contained what he called 'Merchantable commodities' – those products that might make a planter rich or 'enterprisers in general, and greatly profit our owne countrey men'. The second focused on 'all the commodities which wee know the countrey by our experience doeth yeld of it selfe for victual, and sustenance of mans life'. The third addressed 'such other commodities besides ... as I shall thinke behoofull' to future planters and 'a briefe description of the nature and maners of the people of the countrey'.[18]

Among the 'merchantable' commodities Harriot included were 'grasse Silke', 'Worme Silke', flax and hemp, 'Roche allum', 'Wapeih' (white clay), Sassafras, Cedar, wine, 'Oyle', animal furs, deer skins, 'Civet cattes', iron, copper, pearls, sweet gum trees, 'Dyes', 'Oade' (for use by cloth dyers), sugar, and especially pine trees that produced great quantities of 'Pitch, Tarre, Rozen and Turpentine' (needed for making ships.) In the case of each item, he

17 Quinn (ed.), *Roanoke voyages*, vol. 1, 190-91 (*Tyger* journal); 286–87 (this is in Lane's *Discourse*, where Quinn identifies the 'Colonel of the Chesepians' mentioned by Lane as perhaps Harriot). In another reference by Lane to Harriot (vol. 1, pp. 273–74), Quinn suggests that Lane means that he took Harriot's own geographic information and not that Harriot was along on a dangerous trek up the Roanoke River. Harriot's misgivings about English belligerence toward Algonquin Indians appear at the beginning of his discussion 'of the nature and manners of the people' and at the end of his 1588 tract *Briefe and true report on the new found land of Virginia*, in Quinn (ed.), *Roanoke voyages*, vol. 1, 368, 381–82.
18 Ibid., vol. 1, 321, 324–25.

carefully included Indian uses of the item; how it compared with similar or known products in Europe or other nations; and how each item could be produced and sold in the world market.[19]

Harriot expounded more on the abundance of foods in this corner of America for the 'sustenance of mans life', but more importantly 'usually fed upon by the naturall inhabitants: as also by us, during the time of our aboade'. First among these was *pagatowr* (maize or corn) the principal food crop of the Americas. But also *okindgíer* (beans), *wickonzówr* (peas), *macócqwer* (melons and gourds), *melden* (spinach and beets), and *planta Solis* (sunflower). As a close observer of the art and science of producing a rich yield, Harriot provided a detailed description of Indian methods of planting and harvesting these crops. And of the curious but popular 'herb' called *uppówoc* (tobacco) by the Indians. He also included separate categories of root foods, including *openauk* (potato) and *okeepenauk* (yams), among others; and of fruits and nuts, among them chestnuts, walnuts, medlars (persimmon), and especially *metaquesónnauk* (pear) and five varieties of acorns. Under the heading of edible foods, Harriot included beasts, even though some of them – like deer and conies (rabbits) – whose skins were also exported for clothing; 'Foule' such as 'turkie cockes', 'turkie hennes', and geese used both for food and for feathers; and 'Fishe' of many varieties and types of 'crustie shel fishe' complete with the Indians' special methods of fishing 'with poles made sharpe at one ende, by shooting them into the fish after the maner as Irishmen cast dartes; either as they are rowing in their boats or els as they are wading in the shallowes for the purpose'.[20]

The third part of his succinctly chiseled classification included potential products, edibles, and other resources available for exploitation: materials for building homes, making tools, and fabricating watercraft. Oak trees and walnuts for houses; 'firre trees' for ship masts; and *rakíock* trees (tulip or cypress) for making 'their boats or Canoes ... with the helpe of fire, hatchets of stones, and shels ... some so great ... of one tree that they have carried well xx. [20] men at once, besides much baggage'. One thing lacking from the perspective of English peoples were quarries for stone, marble, or limestone to build sturdy homes. But 'in divers places of the countrey there is clay both excellent good, and plenty' for making brick and deep middens of spent 'Oister shels' for making lime 'after the maner as they use in the Iles of Tenet and Shepy, and also in divers other places of England'. Though the 1588 edition of the *Briefe and true report* contained crowded pages of type, Harriot's precise words and comparisons with English practices painted easily grasped mental images.[21]

Harriot addressed 'the nature and manners of the people' in the last portion of his treatise. While the disciplines of anthropology and humanistic

19 Ibid., vol. 1, 325–37.
20 Harriot, *Briefe and true report,* in Quinn (ed.), *Roanoke voyages,* vol, 1, 337–62.
21 Ibid., vol. 1, 362–68.

studies lay centuries in the future, Harriot was one of a few gifted individuals during the English Renaissance, only 27 years of age when he penned his observations, capable of comprehending human character and instincts among a variety of cultures. While others of the Roanoke explorers routinely described the Carolina Indians as *savages* (Hakluyt, Grenville, Lane, and White), Harriot avoided words such as savage, barbarian, heathen, or infidel in his tract. Indeed, Harriot's descriptions of native Indians exhibit an aura of respect sharply at variance with other empire builders in England and Europe. His tone by comparison was carefully nuanced and surprisingly objective.[22]

In his first words about the native Indians, he demonstrated admiration toward them: 'They are a people clothed with loose mantles made of Deere skins, & aprons of the same rounde about their middles; all els naked: of such a difference of statures onely as wee in England'. His non-judgmental tone continued throughout his text: 'Their houses are made of small poles made fast at the tops in rounde forme after the maner as is used in many arbories in our gardens of England'. With respect to their weapons for war, Harriot wrote, 'Although they have no such tools [ordnance], nor any such craftes, sciences and artes as wee; yet in thos things they doe [war], they shewe excellencie of wit'. On topics of theology, 'for mankinde they say a woman was made first' and 'they beleeve also the immortalitie of the soule'. Harriot even apologized for his limited command of the Algonquian language since native Indians wanted 'to learne more then we had means for want of perfect utterance in their language to expresse'. When the English displayed a Bible and talked about its message the Indians wanted 'to touch it, to embrace it, to kisse it, to hold it to their brests and heads, and stroke all over all their bodie with it; to shew their hungrie desire of that knowledge which was spoken of'.[23]

Harriot's esteem for his Indian guests in London and on the Carolina coast was immense. But after his turbulent year of exploring Indian lands and villages along the Carolina inner coast, he was concerned whether he and Englishmen could earn the respect of the Indians they met. His

---

22 The author has conducted a separate analysis of the use of the words savage, barbarian, heathen, and infidel as descriptive terms for both Indians and blacks in colonial America and the early national United States. The word 'savage' appears only once in Harriot's *Briefe and true report*; see Quinn (ed.), *Roanoke voyages*, vol. 1, 333. In that instance, Harriot mentions a collection of pearls assembled by one of the members of the expedition: 'One of our companie; a man of skill in such matters, had gathered together from among the savage people about five thousande [pearls]: of which number he chose so many as made a fayre chaine'. It is not clear why he deviated in this one instance, unless he was paraphrasing what the pearl gatherer had told him or his treatise had been reviewed and perhaps edited by Richard Hakluyt before it went to press.

23 Harriot, *Briefe and true report*, in Quinn (ed.), *Roanoke voyages*, vol. 1, 368–71, 373, 375, 377.

associates had kidnapped and killed Indians, even assassinating the chief *Wiroans* Wingina whose good grace and gifts had enabled them to survive. As he explained near the end of his treatise, 'these their opinions [Indians] I have set downe the more at large, that it may appeare unto you [his readers] that there is good hope they may be brought through discreet dealing and government to the embracing of the trueth, and consequently to honour, obey, feare and love us'. That seemed problematic, given the harsh paternalistic aims and habits of his principal mentors, Ralegh and Hakluyt, to Christianize and manipulate the Indians of North America.[24]

In retrospect, it seems strange that historians should have given Harriot's 1588 treatise such short shrift for over four centuries. Perhaps because Harriot himself acknowledged that it was written in haste to quash rumors that the expedition had been a failure? But within its text he promised there was more to come. 'I have ready in a discourse by it self in maner of a Chronicle according to the course of times and when time shall bee thought convenient shall also be published'. Perhaps because Governor Ralph Lane's separate report noted that on their sudden departure from Roanoke Island many of the journals and records were lost? 'The weather was so boisterous, and the pinnaces so often on ground, that the most of all wee had, with all our Cardes, Bookes and writings, were by the Saylers cast over boord'. Whatever the reason, these disparaging assessments from one generation to another obscured the substantive content of Harriot's treatise, sending a wrongful message that it was bereft of scientific significance.[25]

By misunderstanding and downplaying the historical importance of his *Briefe and true report*, Harriot biographers and historians have also contributed to the marginalization of Harriot as an historical figure. Their repeated insistence on Harriot's failure to publish his work has further eroded his pivotal influence as an author. On the contrary, Harriot was indefatigable in his campaign to publish his *Briefe and true report* not only as an interim report on what he and his colleagues had accomplished in their well-publicized venture, but also as the foundation for a ground-breaking illustrated book on an indigenous people and their habitats for all of Renaissance Europe. Fortunately, his role in making this happen has emerged in recent years through non-Harriot scholars who have demonstrated the timeless importance of his treatise.

We should have understood all along that Harriot had large plans for his report on Virginia. He told us so in the first edition. After listing the 'Foule' of Virginia, he hinted at his grand plan: 'Of al sorts of fowle I have the names in the countrie language of fourscore and six ... [which] we have taken, eaten, & have the pictures as they were there drawne with the names of the inhabitaunts of several strange sorts of water foule eight, and seventeene kinds more of land

---

24 Ibid., vol. 1, 377, 381.
25 Ibid., vol. 1, 322, 387. Ralph Lane, 'Discourse on the First Colony', in Quinn (ed.), *Roanoke voyages*, vol. 1, 293.

foul, although wee have seene and eaten of many more, which for want of leisure there for the purpose coulde not bee pictured'. This disclosure of a growing catalogue of American birds complete with pictures was followed with another cryptic note: 'After wee are better furnished and stored upon further discovery, with their strange beastes, fishe, trees, plants, and herbes, they shall be also published'. If this be combined with his concluding claim, 'I have ready in a discourse by it self in maner of a Chronicle ... [that] when time shall bee thought convenient shall be also published', we have the outline of his projected master-work on Virginia.[26]

Harriot wrote these words in February 1588, three months before word arrived in London that a Spanish fleet of 130 armored ships had left Spain headed toward England, and five months before the Spanish armada and a hastily assembled flotilla of English ships collided off the coasts of England and Ireland. This colossal national emergency not only impeded John White from securing the supplies he needed to save Ralegh's last American colony; it also modified Harriot's plans for publishing his illustrated edition of *Briefe and true report* that would contain both John White's sketches and his new narrative (the 'Chronicle'). Since Ralegh, Grenville, and the other great warriors of England were focused on a naval war with Spain, our hero Harriot, with the assistance of Richard Hakluyt and the concurrence of Ralegh, projected a new and even more ambitious plan for his illustrated edition of *Briefe and true report*.[27]

This is where an exiled French Huguenot Theodor de Bry (1528–98), a master goldsmith and would-be book publisher, entered the scene. During a safe haven from French oppression in London from 1585 to 1588, de Bry decided to apply his artistic skills to the novel art of copperplate engraving. In 1588, he moved to Frankfurt, Germany, the reigning book capital of Europe. As a devout Protestant, de Bry conceived the idea of publishing a series of illustrated books on European (especially Protestant) expeditions to the Americas. The initial choice of art for his first book was the beautifully colored sketches of another French Huguenot exile in London, the artist Jacques Le Moyne (1533–88). Like John White, Le Moyne had accompanied a failed French expedition to Florida in 1564 and produced a superb set of watercolors on the lives and wars of native Indians encountered by the French explorers. Fortunately, Le Moyne resided in the same London neighborhood where de Bry lived in exile. Although Le Moyne was reluctant to sell his drawings at first, de Bry eventually got the art from Le Moyne's widow on a return trip from Frankfurt.[28]

---

26 Harriot, *Briefe and true report,* in Quinn (ed.), *Roanoke voyages,* vol. 1, 358–59, 387.
27 After developing the narrative presented here, I discovered that the great authority on the Ralegh expeditions to America, David Quinn, had organized and speculated on some of the components of this story in the 1970s. See Quinn, 'Thomas Harriot and the New World', in Shirley (ed.), *Thomas Harriot. Renaissance scientist,* 43–47.
28 Ibid., 46–47.

Meanwhile Hakluyt completed his diplomatic mission in Paris and returned to London in 1588 to begin work on the first edition of his own landmark publication, *The principal navigations, voiages and discoveries of the English nation* (1589). Though his breathtaking book contained no illustrations, Hakluyt did incorporate all of the existing narratives of Ralegh's explorations including a reprint of Harriot's *Briefe and true report* with annotations no doubt provided by Harriot. A sudden storm of opportunity occurred to Hakluyt, Harriot, and perhaps also Ralegh. If de Bry could be persuaded to begin his new series of illustrated books with Harriot's narrative and John White's art, the aspirations of England to create a colonial empire could be announced to all of Europe. Although Ralegh was at war, the moment was even more perfect for these English schemers, since at that very moment, Ralegh had an entirely new colony in Virginia building the 'Citie of Ralegh' and the very artist himself John White was its governor. Although de Bry had Le Moyne's artwork in hand and the French expedition preceded the English exploration of Harriot and White by 20 years, the sales pitch of Hakluyt and Harriot carried the day: de Bry's first volume of his 'Grands Voyages' began with Harriot's narrative and White's art.[29] As de Bry explained in a note to readers on the odd sequencing of the later work first, 'I here sett out in the first place [the Virginia book], being thereunto requested by my Friends, by Raeson of the memorye of the fresh and laue [late?] performance thereof'.[30]

Although Harriot's original plan was to produce a new and expanded work as a companion to the White drawings, de Bry decided to stick with Harriot's 1588 print edition of *Briefe and true report*, warts and all, with no textual changes. Thus, what Harriot composed in haste, as historians often point out, became the permanent, authoritative text for his magnificent book. John White's drawings served as sources for the illustrated portions of the book. De Bry acknowledged this decision in his flowery, yet barely grammatical dedication of the English version to Ralegh. 'I have thincke that I cold faynde noe better occasion to declare it [his appreciation], then taking the paines to cott in copper (the most diligentye and well that wear in my possible to doe) the Figures which do levelye represent the forme and maner of the Inhabitants of the same countrye [Virginia] with theirs ceremonies, sollemne, feastes, and the manner and situation of their Townes, or Villages. Moreover', he continued, 'I have thincke that the aforesaid figures wear of greater commendation, if somme Historie ... weare Joyned with the

29 Authors have chosen various ways of narrating this story. But the narrative presented here seems more likely to be correct due to the rich archival sources mined by the Dutch scholar Michiel van Groesen and documented by him in his 'de Bry bible' (my words), M. van Groesen, *The representations of the overseas world in the De Bry collection of voyages (1590–1634)* (Leiden: Brill, 2008), 112–13.
30 T. Harriot, *A briefe and true report of the new found land of Virginia. The complete 1590 Theodor de Bry edition*, ed. Paul Hulton, English version (New York: Dover, 1972), 41.

same ... [such as] the rapport which Thomas Hariot hath lattely sett foorth, and have cause them booth togither to be printed ...'.[31]

What de Bry originally had in mind, conversely, was a Latin edition of Harriot's treatise with engravings of White's images. To assist in that goal, several eager and supportive Englishmen produced four separate Latin translations of Harriot's text to achieve an authoritative Latin text. The volunteer translators included a London-based pharmacist, James Garet; a botanist in the English diplomatic service, Richard Garth; another unidentified correspondent named Francis Rogers; and, of course, Hakluyt himself. De Bry assigned responsibility for decisions on the Latin version to Carolus Clusius (Charles de l'Écluse, 1526–1609), a renowned contemporary horticulturalist and versatile linguist. Despite multiple threads and contributions, de Bry assigned full credit for the ultimate Latin translation to Clusius.[32]

De Bry also gave credit to Hakluyt for his role in conceptualizing the project as a whole. Hakluyt 'first Incouraged me to publish the Worke', de Bry wrote in his preface to the English edition. In addition to persuading de Bry to launch his 'Grands Voyages' with Harriot's treatise and White's watercolors, Hakluyt convinced him to publish the book in four languages. Not only Latin, but also the three major European spoken languages of German, French, and especially English. While de Bry's goal was to sell books, Hakluyt's goal was to announce the emergence of the English empire to as many people (at least in Protestant nations) as possible. De Bry also credited Hakluyt for bringing Harriot's book and John White's drawings to his attention: 'I was verye willing to offer unto you ["gentle reader"] the Pictures of those people [Virginia Algonquians] wich by the helse [help] of Maister Richard Hakluyt of Oxford Minister of Gods Word ... I creaved out of the verye original of Maister Jhon White an English paynter who was sent into the countrye ... to draw the description of the place, ... [and] to describe the shapes of the Inhabitants their apparell, manners of Livinge, and fashions'.[33]

Since Hakluyt was the principal promoter who dealt directly with both de Bry and his language editor Clusius, these credits are understandable. But as de Bry and his sons Johan Theodore and Thomas were busy transforming White's images into copperplate engravings, they may have been unaware that their resultant engravings were in the process of being described by Thomas Harriot, the same person who authored the treatise in the first half of their eventual masterpiece.[34] We know that Harriot wrote the texts for the

---

31 Ibid., 4.
32 Van Groesen, *Representations of the overseas world*, 112–14, including 113, n. 18.
33 Harriot, *Briefe and true report* (Dover edn., 1972), 41.
34 The de Brys focused on the engravings and had little or nothing to do with the translations according to Van Groesen, *Representations of the overseas world*, 121–22. David Quinn documented Harriot's authorship of the Latin texts for the 1590 editions in Quinn (ed.), *Roanoke voyages*, vol. 1, 390, n. 4; 401, n. 1; 414, n. 5; and 430, n. 4. See also Charles Fantazzi, 'Harriot's Latin', in Fox (ed.), *Thomas Harriot and his world*, 231–36.

engravings and that he wrote them in Latin. We also know that, oddly, Hakluyt translated Harriot's texts into English for the English edition. We know that Clusius translated Harriot's Latin texts into French and German. But we do not know how de Bry and Clusius coordinated the creation and circulation of draft engravings for Harriot to write and perfect the Latin texts. But we can determine by the content and quality of the texts that they could only have been written by Harriot himself. Given the extensive correspondence between English Latinists and both de Bry and Clusius, communication and travel between the two places must have been regular and efficient.[35]

By December 1589, Clusius had the engraving translations in hand. De Bry was thus ready to design pages and set type in four languages. Since the same engravings were used in all four language versions, variations in type size and language lengths was a challenge for printers. Since de Bry also wanted to release all four versions simultaneously during the first week of April 1590, the publication of Harriot's epochal book was a Herculean feat of time co-ordination and management. Although the challenge was large and quality control complicated, all four variations appeared on time and Thomas Harriot's dream of a large illustrated book with his texts and engravings inspired by John White's sketches became a reality in the spring of 1590.

We now know with confidence that it was Harriot who wrote the entire basic text for all parts of the 1590 de Bry edition of the *Briefe and true report*. But there are some remaining quandaries about the final content and shape of the engravings that beautifully illuminate the elegant book. Many historians of the Roanoke voyages have wrongly assumed that John White must have played an important role in the creation of the book. But other than supplying a set of his Virginia watercolors to be forwarded to de Bry and his engravers as a basis for the new book's illustrations, no other connection with the resulting book can be detected from currently available evidence.[36]

The de Bry engravings for the *Briefe and true report* departed dramatically from White's original sketches. The most fundamental modifications are these: (1) The engravers introduced a full and abundant landscape for nearly all of the human figures depicted, whereas White's subjects virtually floated in space. Except for White's dramatic fishing scene and two Indian villages, the landscapes were imaginary creations of the engravers[37]; (2) The engravers introduced images of agricultural plants (pumpkins, gourds, sunflowers, and flowers in general) not depicted in White's drawings[38]; (3) de

35 Van Groesen, *Representations of the overseas world,* 114–15.

36 This is essentially the conclusion also of Van Groesen, *Representations of the overseas* world, 112, and David Quinn, 'Thomas Harriot and the New World', in Shirley (ed.), *Thomas Harriot. Renaissance scientist*, 46–47.

37 Harriot, *Briefe and true report* (Dover edn., 1972), 46–54, 62–63, 66–67 (inclusive of Plates II-XI, XVII, XIX).

38 Ibid., 68–69 (Plate XX).

Bry's engravers transformed nearly all of White's human figures from re-latively lean and expressionless forms into muscular, robust and expressive beings;[39] and (4) the engravers copied and pasted items from White's other sketches (fish, turtles, birds) to embellish some engravings. Yet, in spite of these modifications Harriot's texts seem to fit perfectly with what is depicted in each engraving.[40]

This is perhaps most notable in the engravings whose captions include a first-person voice where there was no prototype image among White's drawings. 'Arrival of the Englishmen in Virginia' (Plate II) is a new map of Roanoke Island and the barrier islands with both visual and technical in-formation not in White's prototypes. In this caption while describing the difficulty of navigating the coast, Harriot wrote: 'such was our arrival into the part of the world, which we call Virginia ... wich people, theyr attire, and maneer of lyvinge, their feasts, and blankets, I will particullerly declare unto yow'. Another example is 'Their sitting at meate' (Plate XVI), a basic White depiction, but where Harriot noted 'their meate is Mayz sodden, in suche sorte as I described it in the former treatise of very good taste, deers flesche, or of some other beaste, and fishe'. In another newly created engraving, 'The Marckes of sundrye of the Cheif mene of Virginia' (Plate XXIII), the en-gravers depicted a variety of unique tattoos 'rased on their backs, wherby yt may be known what Princes subjects they bee'. Harriot wrote 'the marks which I observed amonge them, are here put downe', with a list of notes indicating the particular Indian group designated by each design.[41]

The engravings and their captions in the 1590 edition of *Briefe and true report* seem thus to be essentially the product of a distant collaboration between Thomas Harriot and Theodor de Bry's engravers with a touch of editorial oversight by Richard Hakluyt. From the first publication of the book in four languages in 1590 and for the next 400 years, most authors and readers believed that the engravings were accurate replications of John White's artwork. But they were, in actuality, highly modified representations of the inhabitants of Virginia that became influential stylized images of how engravers thought American Indians looked and lived. We know this to be the case because from the moment the 1590 edition appeared, authors and illustrators seized the published images (the images were the same in all four language versions) for use in other publications and works of art.[42]

---

39 E.g., Ibid., 46, 47, 50, 51, 61 (Plates III, IIII, VII, VIII, XVI).

40 E.g., Ibid., 56–57, 61, 71, 72–73 (Plates XIII, XVI, XXI, XXII).

41 Ibid., 45, 61, 74 (Plates II, XVI, XXIII).

42 By 1591, the Italian designer Pietro Bertelli (fl. 1580–1616) began publishing images drawn from the de Bry engravings of Indians in Virginia. A German artist depicted a parade in Stuttgart, Germany in 1600 with de Bry images. Virginia's early Governor John Smith (1580–1631), in his 1612 *Map of Virginia* and his 1624 *Generall historie of Virginia*, re-plicated the de Bry images liberally and freely. So did Governor Robert Beverley in his *History and present state of Virginia* (1705). So too did North Carolina textbook authors

The use of the de Bry Indian images as design elements in maps was universal. For example, in 1593 in one of the earliest maps of North America, the Antwerp mapmaker Cornelis de Jode copied six of the Indian figures (colored) for his *America pars Borealis, Florida, baccalaos, Canada, corterealis*. In 1606 Jodocus Hondius in Amsterdam issued a new edition of Gerhard Mercator's pioneering world atlas by adding a new map specifically of the region from Virginia to Florida inserting a version of the de Bry engravings of the village of Pomeiooc and of 'Their manner of fishynge'.[43]

One of the reasons why the Harriot-de Bry images could be so easily copied and adapted to a variety of settings was that the books were widely available, especially in engravers' studios. Additionally, until the second half of the nineteenth century, a gentleman collector or a library could secure Latin and German copies of the Harriot-de Bry volume quite readily. The number of copies that could be printed was limited to some extent by the useful life of the original copper plates. By wear and tear, the plates began to lose quality after a thousand prints. Although English and French versions were limited to the Harriot volume, second and third editions of the German version were issued in 1600 and 1620. A second Latin edition was released in 1608.[44]

During the nineteenth century, one bookdealer, Henry Stevens, Jr., and his son Henry N. Stevens were able to find a sufficient abundance of Harriot-de Bry copies and partial copies around Europe to operate a robust business of selling 'perfect' copies (i.e., assembled from damaged books) to collectors and libraries in England and the United States. Stevens and his son cornered a near monopoly in the sale of such books from the 1850s through the 1920s, gloating on how they struck this rich vein in the international book market.

well into the twentieth century. Even the de Brys themselves pulled out their Harriot volume engravings to print a poster in 1617 titled *Nine Virginia indians and a child*. These replications are depicted in separate essays by Christian F. Feest and Ute Kuhlemann in Kim Sloan's brilliant catalogue of John White's art, *A new world. England's first view of America* (Chapel Hill, NC, 2007), 64–64, 77, 89–92. See also Van Groesen, *Representations of the overseas world*, 357–61.

43 W. C. Wooldridge, *Mapping Virginia from the age of exploration to the Civil War* (Charlottesville, VA, 2012), 18, 22, 39–40, 45, 53, 341–43 and Van Groesen, *Representations of the overseas world*, 357–61. Other examples are John Smith's 1612 map, which included modified versions of the 'Tombe of their Werowans' and of 'A great Lorde of Virginia' holding an axe or club in his left hand. A similar theme was repeated in thirty-seven editions of Hondius atlases, with the axe moved to the right hand. Hondius licensed use of the image in 1618 to Willem Blau, who also used the Smith decorations but returned the great lord's axe back to the left hand. A new interpretation of this map was created by the English mapmaker Ralph Hall in London, with numerous new decorations including an emended version of 'Their manner of prainge', the palisade of Pomeiooc with 'Their dances' inside, and the tomb of chiefs engraving. This practice continued in maps of Virginia and North America through the 1670s and in decorative maps through the eighteenth century and beyond.

44 Van Groesen, *Representations of the overseas world*, 132–33, 390–91, 413, 430, 471.

While one might look askance at this strategy, they distributed Harriot-de Bry books widely across the United States.[45]

A side-effect of stockpiling Harriot-de Bry books by skilled book dealers was the escalation of prices and a near halt in their distribution in the twentieth century. The dissolution of many private and public libraries in Europe during the Second World War expanded the supply of Harriot-de Bry books and other volumes of Theodor de Bry's 'Grands Voyages' temporarily. But by the opening decades of the twenty-first century, the antiquarian book market had recovered sufficiently that individual copies of Harriot-de Bry books – even mongrelized 'perfect' copies – sold for as much as $100,000 each. Hand-colored versions cost much more.[46]

Meanwhile, other editors and publishers began producing quality reprint editions that kept Harriot's invaluable book available to a wider audience. Some of these editions coincided with the 300th anniversary of the Roanoke voyages (1884–90) amid a resurgence of interest in the origins of English America. The renowned bibliographer of early America, Joseph Sabin (1821–81) issued a photo-lithographic facsimile reprint from New York in 1871.[47] The Holbein Society, with the English antiquarian W. H. Rylands as editor, produced another facsimile reprint in 1888 from Manchester.[48] The distinguished bibliographer and book dealer Bernard Quaritch (1819–99) published his own facsimile reprint in 1893.[49]

By the second half of the twentieth century, a very inexpensive quality reprint of the English version of Harriot-de Bry came on the market. Henceforth Harriot's great book could be assigned for reading in history, art, American history, and Indian studies classes across the English-speaking world. Produced as an initiative of the Rosenwald Collection at the Library of Congress, the Dover Publications edition of Harriot's *Briefe and true report* made Thomas Harriot's name and his book known to average college students throughout the United States and across the world.

One of the ironies in the history of Harriot's book is that its engravings have always been identified as accurate representations of John White's original sketches made during his visit to the Carolina coast in 1585. But from 1590, when Harriot's illustrated book was published, almost no one on

45 L. E. Tise, 'The "perfect" Harriot/de Bry: cautionary notes on identifying an authentic copy of the de Bry edition of Thomas Harriot's *A briefe and true report* (1590)', in Fox (ed.), *Thomas Harriot and his world*, 216–29; H. Stevens, *Recollections of James Lenox and the formation of his library* (New York, 1951), 56–60, 110, 113–14.

46 Tise, 'The perfect Harriot/de Bry', in Fox (ed.), *Thomas Harriot and his world*, 201–17.

47 T. Hariot, *A briefe and true report of the new found land of Virginia ...* (New York, 1871); OCLC 5854190.

48 T. Hariot, *A briefe and true report of the new found land of Virginia ...* (Manchester, 1888); OCLC 8574314.

49 T. Hariot, *Narrative of the first English plantation of Virginia* (London, 1893); OCLC 1618054.

earth (other than the book dealer Henry Stevens) could have known that there were dramatic and essential differences between White's drawings and the engravings. Since Stevens acquired the White originals at a Sotheby's auction on 11 August 1865 (after they had been charred in a fire and then water-soaked in a Sotheby's warehouse), he could have explained to one and all that there were fundamental differences between the watercolors and the engravings. But he did not. Unable to find an American buyer for the White originals, Stevens sold them to the British Museum in March 1866, after cleaning and remounting them. This was not one of his most lucrative sales. His winning auction price was £125. The Museum paid him £236. And there they have remained from 1866 to the present.[50]

Despite their presence in the greatest repository of the artifacts of British history, the collection remained unknown until 1905, when they were moved from the Museum's Department of Manuscripts to its Department of Prints and Drawings. Although the collection of 75 watercolors was then listed in the Museum's authoritative *Catalogue of drawings by British artists*, they remained largely unknown and unseen except to the Museum's curators. A few Americans visited the Museum and made semi-clandestine black and white photographs or hand-colored copies of some of the images for their personal research purposes.[51]

A concerted effort by librarians at the Huntington Library, the William Clements Library, the Library of Congress, and the British Museum in the 1930s to print copies of the original White drawings came to naught. However, Randolph G. Adams of the Clements Library was not to be denied. During a visit to London in the early days of the Second World War, Adams, determined to preserve the White images before they could be destroyed by German bombs, arranged to make faint photostats of the watercolors on drawing paper and to have these hand-colored in the presence of the originals. During the war, Adams also entered into collaboration with a flamboyant Hungarian-born filmmaker and popular author Stefan Lorant (1901–97) to publish these hand-colored copies along with a set of the Harriot-de Bry engravings. The book titled *The new world: the first pictures of America*, published in 1946, was a sensational success and introduced many Americans for the first time to a juxtaposition of de Bry's engravings alongside only an approximate idea of the complex colors of the White

---

50 Sloan, *New world*, 94–95. Stevens, *Recollections of James Lenox,* 112, 114–15, n. 10.

51 David I. Bushnell, Jr. (1875–1941), a young amateur anthropologist from a wealthy St Louis family, traveled with his mother to Europe between 1904 and 1906 in search of antiquities relating to American Indians. In the process, he visited the British Museum, making black and white photographs of the John White drawings. These images were published in two lengthy articles appearing in the *Virginia magazine of history and biography* in 1927 and 1928. See J. R. Swanton, 'David I. Bushnell, Jr. (1942)', in *AnthroSource*, accessed 1 September 2020. https://anthrosource.onlinelibrary.wiley.com/doi/pdf/10.1525/aa.1942.44.1.02a00100omplex.

originals. Despite the poor quality of the White reproductions, this book – a perennial offering of America's Book of the Month Club – made its way into numerous homes and became a favorite coffee table curiosity.[52]

A similar procedure was adopted in 1964 by the British Museum in a collaboration with the University of North Carolina Press to publish an elegant two-volume elephantine edition of the White watercolors. Titled *The American drawings of John White, 1577–1590,* this lavishly funded project included an introductory volume with a history of the collection and of each watercolor by the two greatest authorities on White and his art: David Quinn and Paul Hulton, Assistant Keeper of Prints and Drawings at the British Museum. New hand-colored reproductions of the White originals were 'executed in collotype and *pochoir* (stencil)'. The collotypes were 'finished by using finely cut metal foil stencils, through which water-colour is brushed on by hand—a combination of processes ... for a long time ... employed with success in France'. In fact, the reproductions were finished in Paris. Despite the extravagant cost and elaborate process, the end products were still but approximations of the White originals.[53]

In 1984 the British Museum for the first time authorized photographic reproductions of the White watercolors for the 400th anniversary of the first Roanoke voyages of 1584. The Museum additionally put some of the images on loan to selected American museums. Paul Hulton wrote a text and captions for a catalogue accompanying the traveling exhibition titled *America 1585: the complete drawings of John White* (1984). In this delayed and contorted process, Americans got their first glimpse of White's original watercolors of North American Indians.[54]

Just a few years later, in 1992, the world saw for the first time, at least in France, a book that reproduced one of the finest hand-colored versions of the de Bry edition of Thomas Harriot's book. The book, under the imprint of the prominent French publisher Gallimard, reproduced the hand-colored engravings of the first three volumes of de Bry's 'Grands Voyages'. In copies preserved in the collections of the Bibliothèque du Service Historique de la Marine at the Château de Vincennes, the hand-colorist used a sensuous palette of rich colors to interpret the settings, scenes, and Carolina coast Indians in brilliant hues far more luminous than John White's original pale watercolors.[55]

---

52 S. Lorant, *The new world. The first pictures of America,* revised edition (New York, 1965), esp. 180–84. This was, in fact, the book, with its 1965 modest revisions, that introduced me to the worlds of Thomas Harriot and John White in 1967 while I was a student and an avid buyer of Book of the Month Club offerings.

53 P. Hulton and D. B. Quinn, *The American drawings of John White, 1577–1590,* 2 vols. (Chapel Hill, NC, 1964), vol. 1, xii-xiii, 58–61.

54 P. Hulton, *America 1585. The complete drawings of John White* (Chapel Hill, NC, 1984).

55 M. Bouyer and J. P. Duviols, *Le Théâtre du nouveau monde. Les Grands voyages de Théodore de Bry* (Paris, 1992), with the Harriot engravings and captions, 2–47 and a commentary on each engraving, 129–56.

But the real heyday in the recognition of the 1590 edition of Harriot's *Briefe and true report* as one of the world's most significant books began in the first decades of the twenty-first century. Two landmarks in the study and appreciation of Harriot's achievement occurred coterminous with the 400th anniversary of the founding of the first permanent English colony in North America at Jamestown, Virginia. The first of these was the publication for the first time of high-density scanned images of all of John White's watercolors in the British Museum alongside scanned black and white images of corresponding engravings in an English version of Harriot's book in the British Library. The White watercolors appeared with corresponding de Bry engravings in a brilliant catalogue and exhibit organized by Kim Sloan, Curator of British Drawings and Watercolours at the British Museum. Years in preparation and distilling the research of dozens of curators, historians, cartographers, anthropologists, art historians, and many more, *A new world: England's first view of America* presented 400 years of authoritative information on both White's watercolors and Harriot's achievements. While the book was designed as an exhibit catalogue, it will remain a compelling testament to White's newly revealed watercolors and their reconfiguration in the engravings of Harriot's book.[56]

A second landmark publication appeared a year later. Michiel van Groesen's *The Representations of the overseas world in the De Bry collection of voyages (1590–1634)* will be referenced mainly by specialists. However, van Groesen, a Dutch historian, provided a great service to anyone interested in grasping the vast maze of Harriot-de Bry translators, contributors, engravers, printers, and booksellers involved with the book and other volumes in de Bry's 'Grands Voyages'. From the perspective of collectors, dealers, librarians, curators, and historians, van Groesen also created a map for navigating the many editions, states, reproductions, and formats in which Harriot's book appeared in the sixteenth and seventeenth centuries. After years of prodigious research, van Groesen distilled mountains of original data on Renaissance publishing into a mere 600 pages (!) that will assist researchers wend their way through the thicket that created and promoted Harriot's great book.[57]

These monumental publications have facilitated numerous research projects relating to Thomas Harriot and his book. One partially previewed in Kim Sloan's catalogue, albeit in a footnote, reported preliminary findings of new research project, 'a census of the 1590 editions of the Harriot/de Bry—focusing on coloured versions'. The survey was 'being conducted by the American historian Larry E. Tise at the newly named Harriot College of Arts and Sciences at East Carolina University'. The note continued: 'Preliminary

---

56 Sloan, *New world,* including J. E. Chaplin, C. F. Feest, and U. Kuhlemann as other principal contributors.
57 Van Groesen, *Representations of the overseas world.*

findings in Tise's study [2007] suggest that de Bry's book has been bound and rebound in many creative ways. Those seen to date in libraries in the USA, France, and Britain are highly varied in colours, colour schemes and applications used for different languages ... and tended to reflect distinct cultural perceptions or interpretations of American Indians'. Not a bad summary of the nature of the research and the ultimate findings of the project.[58]

Figure 7.2 Theodore de Bry's *India Occidentalis* or "Grands Voyages" in the Anne S. K. Brown Military Collection, John Hay Library, Brown University. This remarkable set of Theodor de Bry's "Grands Voyages" was first encountered by the author in January 2007. 6 volumes, in Latin, hand-colored, with original or early bindings. And a color scheme that used ample amounts of gold to accentuate nearly all engravings in the books. Due to the fact that each volume has been catalogued separately since acquisition by the John Hay Library, the existence of this unique hand-colored version of the "Grands Voyages" was unknown beyond the library itself. This one-of-a kind hand-colored set of books became the basis for *Theodore de Bry—America: The Complete Plates, 1590–1602* (2019). Photo taken by the author October 2016.

---

58 U. Kuhlemann, 'Between reproduction, invention and propaganda: Theodor de Bry's engravings after John White', in Sloan, *New world,* 79–92, 245–46, n. 41.

Figure 7.3 The hand-colored title page of *Admiranda narratio*, Latin version of Thomas Harriot's *A briefe and true report of the new found land of Virginia* published by Theodor de Bry at Frankfurt, Germany in 1590 (shown here in black and white). Courtesy of Anne S. K. Brown Military Collection, Brown University Library. From the John Hay Library's 6 volume hand-colored set of Theodore de Bry's "Grands Voyages." Harriot's book as the inaugural volume for the "Grands Voyages" series was published in Latin, English, French, and German and was dedicated to Sir Walter Ralegh. The title page for each linguistic variation contained the same artistic engraving—but the text panel in the center of the page was modified to reflect its language version.

The census continued for another ten years as more hand-colored copies emerged in the four-language versions of the 1590 Harriot edition in the United States, France, Germany, Belgium, and the Netherlands. Eventually Tise declared that he had seen and analyzed 'all known and accessible' copies in libraries, museums, and private hands. Among the greatest treasures encountered quite unexpectedly was a six-volume hand-colored set of de Bry's 'Grands Voyages' at the John Hay Library at Brown University in the United States. This impressive Latin collection in their original or early bindings (see Figures 7.2 and 7.3) had been colored by a single hand from beginning to end. With this discovery, the goal of the project was transformed from census to the publication of the Hay Library's extraordinary hand-colored volumes, including, of course, a history and analysis of hand-colored versions of Harriot's great book.[59]

Fortunately for Tise's revised goal of publishing this unique hand-colored collection – and for creating another landmark in recognizing Thomas Harriot's magnificent book – the world's most prominent publisher of quality art books arrived on the scene. In a serendipitous moment of fortune and opportunity, the international art publishing house of Taschen provided a vehicle for presenting Harriot's book and nine volumes of de Bry's 'Grands Voyages' in a beautifully designed book worthy of their eminent stature in the history of exploration literature. The unanticipated prospect of uniting the unrivaled scholarship of Michiel van Groesen on the creation of de Bry's 'Grands Voyages', Tise's research on hand-colored versions of de Bry's books, and the book design and marketing capabilities of Taschen was a dream come true. Although the period of time from conception until delivery took another six or seven years, the end result far exceeded everyone's fondest dreams.

The stunning book titled *Theodore de Bry—America: the complete plates, 1590–1602* appeared in 2019 with van Groesen as editor and translator and Tise as authority on hand-coloring and hand-colored editions. Both van Groesen and Tise authored detailed introductions on their areas of expertise. Taschen created a sophisticated and graceful vehicle with cloth-printed covers and linen pages. The elephantine size of the book (11.2 ×15.6") and 376 pages contained an extraordinary amount of material: volumes 1–9 of de Bry's 'Grands Voyages'; reproductions of 300 hand-colored images; and the first-time translations of volumes 2–9 into both English and French. The book included all six of the beautifully hand-colored volumes from the Hay Library. But perhaps most remarkable of all, the book was published simultaneously in three languages: French, German, and English. A fitting reminiscence of the

59 This declaration of having seen and analyzed 'all known and available copies' of hand-colored Harriot-de Bry volumes was first enunciated at a conference titled 'Paint over print' at the Kislak Center of the University of Pennsylvania in February 2015. But Tise has made this claim other times in lectures and talks in the United States and abroad, principally to elicit comments about other copies that may have been overlooked or previously unknown. It should be noted that some private owners of copies and book dealers have declined to have their copies analyzed; Larry E. Tise and Chet Van Duzer, organizers, 'Paint over print: hand-colored books and maps of the early modern period', 19–20 February 2015, Kislak Center, University of Pennsylvania. See also Larry E. Tise, 'America's first "coloring book": Theodor de Bry's 1590 edition of Thomas Harriot's *Briefe & true report of the new found land of Virginia*' https://www.youtube.com/watch?v=QXQqQp6CwTg

simultaneous publication of Harriot's original book of 1590 in four languages. And since the book was also published as a trade book (not a collector's limited edition), it is certain to extend even further the renown of Thomas Harriot and his magnificent book for generations to come.[60]

Figure 7.4 Plate XVI, "Their sitting at meate" (German). For purposes of examining and interpreting color schemes in language variations of Harriot's book, Plate XVI or "Their sitting at meate" was chosen for analytical focus. This image (shown here in black and white) contains most of the color elements where one is likely to encounter color variations: bodily skin, hair, and tattoos; clothing, jewelry, and adornments such as feathers; typical daily needs as food, gourds (for water), tobacco pouch, nuts, fish, corn, and shells; but also landscape, sky, and clouds. In the hand-colored copy of this German print the woman's body is white; her hair is blonde, her eyes blue, and with red lipstick—very unlikely colors to be encountered among American Indians. From Thomas Harriot, *Wunderbarliche, doch warhafftige Erklärung vonder Gelegenheit und Sitten der Wilden in Virginia* (1590), North Carolina Collection, University of North Carolina Chapel Hill FVCC970.1 H28w. Courtesy of Wilson Library Special Collections, University of North Carolina.

60 M. van Groesen and L. E. Tise, *Theodore de Bry—America. The complete plates, 1590–1602* (Cologne, 2019).

If North Carolina's special agent Peter Browne from 1820 went in search of a hand-colored copy of Thomas Harriot's map of the Carolina coast today, he would be able to find digital images of that map from a multiplicity of libraries and museums around the world, with color variations inspired by whether the 1590 edition was originally published in English, French, German, or Latin. But each of them bearing the name of its distinguished author Thomas Harriot.

Note 1: Interpreting hand-colored versions of Harriot's book

A Note on Interpreting hand-colored copies of 'Their Sitting at Meate' (Plate XVI, see Figure 7.4) from Theodor de Bry editions of Thomas Harriot, *A briefe and true report of the new found land of Virginia* (1590)

English versions:

> English language hand-colored versions of this image usually contain muted and solid colors for clothing and skin. Also minimal color emphasis or highlighting of hair, eyes, jewelry, tattoos, and such details as food, tobacco pouch, corn, and food in platter. Plain yellow corn may be suggestive that this is flour rather than sweet corn.

French versions:

> French language hand-colored versions of this image tend to focus much more on the bodily features and clothing details of the man and the woman. Skin colors are darkened for both man and woman. Pinkish colors for hair, and much attention to details of color on animal skin clothing. Details of gourd, tobacco pouch, and ears of corn are accentuated. Corn color scheme is a concentric, improbable design, but suggestive of sweet corn as indicated in the text. Aqua and turquoise tones appear in landscape, sky, and food.

German versions:

> German hand-colored versions of this image generally introduce more color, but with darker tones for landscape, sky, and clothing. In these color schemes, the woman is often lighter in skin color and displays blonde hair, blue eyes, and red lips. The tobacco pouch is bright red and the food in the platter and the ears of corn are multi-colored in neat implausible red, yellow, and blue lines.

Latin versions:

> Latin language hand-colored versions of this image tend to have a great variety of color schemes. Some contain royal colors of purple and red.

Treatments are often highly luminous and intensely vivid in all features, including sky, bodies, clothing, and details. The man's hair may be ruby red as suggested in the book's text. Women in these color schemes recede to the background as opposed to French and German versions that make images of women much more prominent. Whether by chance or special knowledge, the ears of corn displayed in random hues of red, white, and blue more closely approximate the natural colors of Indian sweet corn.

Note 2: Accessible print and online hand-colored versions of Harriot's book

Finding hand-colored copies of 'Their Sitting at Meate' (Plate XVI) from Theodor de Bry editions of Thomas Harriot, *A briefe and true report of the new found land of Virginia* (1590)

Print examples:

> *English, French, German, & Latin*: Michiel van Groesen (ed.) and Larry E. Tise. *Theodore de Bry—America: The complete plates 1590–1602* (Cologne: Taschen, 2019), 32, 34–35.

> *Latin*: Marc Bouyer and Jean-Paul Duviols, *Le Théâtre du Nouveau Monde. Les Grands Voyages de Théodore de Bry* (Paris: Gallimard, 1992), 22–23.

> *Latin*: Thomas Hariot, *A briefe and true report of the new found land of Virginia. The 1590 Theodor de Bry Latin edition* (Charlottesville: University of Virginia Press, 2007), 68, 153.

Online examples:

> *English*: 'The first decade conteyning the historie of travell into Virginia Britania'. Manuscript, [1612]. Princeton University Firestone Library Special Collections. https://dpul.princeton.edu/catalog/eac2a7cc9d1c2c 709a35e9a443371fb8, Image 237

> *English*: 'The historie of travaile into Virginia Britannia'. Sloane MS, 1622. British Library. https://www.bl.uk/collection-items/coloured-engravings-of-native-americans-and-pictshttps://www.bl.uk/collection-items/coloured-engravings-of-native-americans-and-picts, Image 23.

> *German*: 'Picturing the New World: The hand-colored de bry engravings of 1590'. Wilson Library Special Collections, University of North Carolina at Chapel Hill https://dc.lib.unc.edu/cdm/singleitem/collection/debry/id/58/rec/18

8   On Writing Harriot's Biography[1]

Robyn Arianrhod

The enigmatic Thomas Harriot gets under your skin: three decades of Oriel Harriot Lectures attest to that. Each scholar has painstakingly analysed a selection of Harriot's terse, fragmented manuscripts or dug deeper into the history of his times, and each has added new colour to the faded tapestry of Harriot's life and legacy. A life that ended 400 years ago, and a legacy that had all but disappeared from history when the first generation of 'Harrioteers' began to resurrect it in the 1950s. When I first discovered this growing body of scholarship about Harriot, I was intrigued. Why does he draw so much effort from an increasing number of researchers? What is it about him that compels so many to want to restore him to his rightful place in history?

One answer is genius, that magical quality that manifests only in a chosen few – and a 'lost genius' is even more tantalizing. Of course, there is also a broader, less romantic answer: filling in the gaps in early modern history, including the history of science and mathematics in Harriot's time, the era that did so much to pave the way for modern science, and which culminated magnificently in Isaac Newton's *Principia*. A biographer, however, also needs to become invested in the actual person behind those broader historiographical aims.

My own interest in Harriot began in a rather casual way. While writing an earlier book, I finally got around to reading Newton's *Principia*: before then, I had simply taken it for granted as the blueprint for modern theoretical physics. To be sure, Newton himself did not claim to be a 'physicist' or even a 'scientist', for these terms only came into widespread use in the nineteenth century; he simply saw himself as a philosopher pondering the Creator's universe. Nevertheless, his work proved so transformative that I began to think further about the culture that produced him. After all, his Continental predecessors are justly famous, especially Johannes Kepler and Galileo Galilei, but his local influences are not so widely known. In *Principia,* he cited his older contemporaries John Wallis and Christopher Wren, but there was surely a longer tradition: the medieval Merton school, for instance, and

1 Robyn Arianrhod, *Thomas Harriot. A life in science* (New York: Oxford University Press, 2019).

DOI: 10.4324/9781003096580-9

early modern scholars such as Robert Recorde, John Dee, William Gilbert, Thomas Digges, Henry Savile, Henry Briggs, and John Napier. Important figures all. But did any of them have both the experimental and mathematical breadth and depth that Newton showed? Then I happened upon Tom Whiteside's assessment of Harriot, of whom I had read only brief references in history of mathematics texts. The 'depth and variety' of Harriot's expertise gave good grounds, Whiteside believed, for claiming him as 'Britain's greatest mathematical scientist before Newton'.[2] I noticed the anachronistic shorthand in the quote, but I got the message.[3] And I was hooked: who *was* this man who never published but who garnered such high praise from the scholar who knew Newton's work better than almost anyone?

I had no idea then what a long and demanding journey I would undertake in search of Harriot, or that my effort to substantiate Whiteside's claim would lead me to the polite but intense debates about Harriot's standing that have appeared in earlier volumes of Harriot Lectures. These debates have often focused on Harriot's achievements compared with those of his famous Continental contemporaries, especially Galileo and Kepler. But there were other areas of contention, too: for instance, between those interested in Harriot's manuscripts for their scientific and mathematical content, and those who reject the very term 'scientific' in connection with Harriot, and who prefer to focus not on his anticipation of later discoveries but on the historical context in which he worked. Then there were the postcolonial debates, about Harriot's role in the attempt by his patron Sir Walter Ralegh to found an English colony in America, on (North Carolina) Algonquian land. In other words, when I embarked on my odyssey with Harriot, I had no idea that this mysterious man would turn out to be almost as controversial as he was fascinating.

But what a journey it has been! First and foremost, I found it a privilege to spend so much time with such a brilliant and intriguing man; a private man, too, who had burned or buried most of his secrets, and who had been so reluctant to publish his breakthrough works that he never did get around to it. As Harriot scholars know, there are many reasons for this. For a start, there were no scientific journals at the time, and the ancient practice of circulating manuscripts to selected readers was still current. But there were more personal reasons, too. These included Harriot's lack of interest in self-promotion, perhaps partly due to his comfortable financial situation, which, in turn, was thanks to his generous patrons; the need to protect his patron

2 D. T. Whiteside, 'Essay review: in search of Thomas Harriot', *History of science*, 13 (1975), 61–70.

3 With due recognition of anachronisms and hindsight – 'scientist' was coined in the nineteenth century, and 'science' did not yet have its modern, dispassionate meaning – for convenience I shall generally use the terms 'scientist' and early modern 'science' here. See also my 'declaration of bias', below.

Ralegh's commercial interests; his perfectionism; the frequent interruptions to his trains of thought because of his patrons' needs and his own bouts of ill-health; and his fear of falling foul of authorities at a time of heightened political insecurity and religious division: as he told Kepler, in connection with his atomism, 'Things are in such a state here that it is not lawful for me to philosophize freely; we are still stuck in the mud'. Had he lived longer than 'the age of sixty or thereabouts',[4] it is quite possible he would have left a published legacy; as it is, his dying wish was that his unpublished mathematical innovations be made available for any 'public uses as it shall be thought convenient by my executor'. Still, I was sensitive to the fact that while mathematical ambition was one thing, Harriot did not intend his whole life and oeuvre to be prodded and probed. If he had done, surely he would have left us more clues!

It was clearly such an interesting life, though – and a significant one for all manner of historians, because it intersected with so many important issues: political, geopolitical, ethnological, colonial, navigational, religious, and cultural, not to mention the gamut of early modern physics and mathematics. So whatever Harriot's own sense of privacy might have been, his life and work have become public property. The best I could do for him was to try to do justice both to his work and to the man behind the marks on those precious, 400-year-old manuscripts. And it was, indeed, his manuscripts that were to form the basis of my approach: they are virtually all that we have in Harriot's hand, and they form the bulk of the available primary material related directly to him.

These manuscripts – around 8,000 pages – are almost entirely devoted to mathematical and physical problems, so this was to be my primary focus, too. I felt that Harriot's research deserved a broader and more accessible treatment than I had seen anywhere in the literature, including John Shirley's landmark and still important 1983 biography. After all, thanks to Shirley's lead and that of the other pioneering Harrioteers, many scholars have now put their energies and skills into unravelling Harriot's laconic manuscripts, and I have drawn gratefully, with due acknowledgement, on the fruits of their labours: understanding Harriot and his multifarious work is far too big a job for one person. (Oxford University's first and somewhat controversial Harriot scholars, Abraham Robertson and Stephen Rigaud, found this out the hard way.[5])

Although my primary focus was on Harriot's science and mathematics, including his American discoveries and preparations, I also wanted to create

---

4 'Sixty or thereabouts' is from Northumberland's memorial plaque to Harriot.

5 Robertson and, three decades later, Rigaud, were tasked with examining Harriot's manuscripts after Zach found them and urged Oxford to publish them. Their preliminary assessments of Harriot's research were inadequate, to say the least. But it was an impossible task for one person at the time. Zach's wish has finally come true with the digitization of the MSS, but it has been an enormous task – undertaken by the team headed by the late Jackie Stedall, Matthias Schemmel, and Robert Goulding – to collate and translate them.

a biographical narrative – as best I could, given the sketchy details of his life. So I needed to offer readers some general historical context, especially as it related to Harriot's work and to the people and dramas in his life. But I did not intend a scholarly historical analysis or overview, nor did I want to overlap too much with Shirley's biography. I just wanted to give enough historical background to bring Harriot to life, and to show him as both a man of his times and an exception – a genius who would probably have made a decisive contribution to scientific history if he had published, but whose work nevertheless adds to our understanding of that history.

This decision not to attempt a broader historical analysis was partly because of my focus on Harriot's work, and partly because I also aimed to bring him to a wider audience. I hoped that by synthesizing and briefly elucidating the whole range of his research – from pure and applied mathematics to astronomy, navigation, optics, and mechanics, as well as linguistics and ethnology – my book might offer a useful overview for new Harriot scholars, or for those whose focus has been on Harriot's life and times rather than his intellectual output. But I was primarily writing for non-specialists, readers who enjoy serious popular science and who would be intrigued to discover this long-lost genius.

So I had set my goals fairly early in the process, and I knew I had many reliable guides, thanks to all those amazing Harriot scholars. But since these guides did not necessarily agree with one another, my next task was to figure out what *I* thought about Harriot and his work. For there can be no ultimate resolution of the debates I mentioned earlier: everyone who approaches Harriot all these centuries later is grounded in, and biased by, their own interests and specialities. Unfortunately, not everyone keeps this in mind when they review their colleagues' work, and I knew that no biographer could hope to bring new insight into every aspect of a life and career as diverse and poorly documented as Harriot's, certainly not in 250-odd pages.[6] And while I hoped that specialists would forgive my simplifications, recognizing that this was a deeply researched but popular-level book, I knew from the outset that I would have my critics. I knew that becoming serious about this book meant undertaking a risky labour of love.

But I believed in Harriot; at least, I believed in the Harriot of my own fledgling conception. To deepen that perception, I needed to spend more time with his manuscripts. Thanks to the efforts of Jackie Stedall, Matthias Schemmel, Robert Goulding, and their team at the Max Planck Institute, digital versions of Harriot's papers were being made available, and they ultimately proved invaluable to me. Initially, though, I felt it was necessary to see the originals, especially if I wanted to feel the presence of Harriot himself.

---

6 250 pages of narrative plus, as it has turned out, almost 100 additional pages of notes, explanatory diagrams, and references (where I wanted both to acknowledge the specialists on whom I have drawn, and to point interested readers to further material).

Since he preserved so very few personal documents, barely even a letter and certainly not a diary, it is telling that he took care to ensure that his manuscripts survived. He was especially concerned for his 'mathematical papers', which he asked his executors to store in 'a convenient trunk with a lock and key' in the library of his second patron, the Earl of Northumberland. How lucky we are that Harriot's disciples placed far more than just his mathematical papers in said trunk, even if they did lie there unnoticed for 163 years after Harriot's death, until Franz Xaver Zach discovered them in 1784, in Northumberland's library at Petworth. We are also indebted to Henry Stevens, who discovered Harriot's will in the 1880s: it shows how highly he regarded his achievements in symbolic algebra. (His bequests also suggest that he was thoughtful and sociable.) Yet the papers he himself preserved were devoted to a much wider range of scientific and mathematical research than algebra – with just a few tantalizing hints of other interests and dramas – and this gives us a measure of what mattered most to him.

When I studied those pages, with all their laborious hand-done calculations and carefully replicated experiments, all their loose ends, dead ends, and ultimate triumphs, I began to learn something of the man behind them. Not only was he focused on science and mathematics; he was meticulous, patient, hardworking, and endlessly curious. He wanted to know exactly how and why a rainbow forms, and how gravity affects falling objects and the paths of projectiles. He wanted to know if there was an experimental law of refraction, and whether or not an atomic theory of matter could explain why light is both reflected and refracted at transparent interfaces. He was fascinated by astronomy and the new celestial world opened up by his custom-made telescopes. He wondered how many people the earth could support, whether it was possible to calculate the length of an infinitely twisting spiral, and how to find new ways of solving algebraic and geometric problems in pure mathematics. He was curious about all this and more, and he toiled away for years, experimenting and calculating, until he found answers to his questions.[7]

I found it a rather poignant experience surveying Harriot's lifetime of unpublished effort. I knew from my own experience as a mathematician how demanding and frustrating it is to carry out page upon page of careful calculations, only to find an error or other stumbling block, but I knew the joy of discovery, too. As I returned to Harriot each day in the British Library's Manuscripts Reading Room, it felt intimate, as if I were visiting a new friend, getting to know what fascinated him, observing how he approached problems, noting the little quirks: the scratchy, sometimes scrawled quill-penned handwriting, the unusual predominance of mathematical symbolism and data, and the relative lack of verbal explanation. When he did use words, I noted the

---

7 Of course, some of Harriot's research was motivated by the practical needs of his patrons, although he often ended up taking these investigations into deeper, more fundamental territory. I discuss this further later in this essay.

rather ungainly English passages among his usual concise Latin. Only rarely, though, did his manuscripts offer a more direct glimpse of the man behind the intellect: a shopping list or two scribbled in a margin; a signature or doodle here and there in the unique phonetic script he had developed to represent the Algonquian language; some attempts at Biblical chronology (quite likely as research for Ralegh's monumental *History of the world*); a few word games; and occasional references to people he knew or books he wanted.

I noticed, too, that some pages were discoloured, especially those dealing with refraction: it was a testament to their importance, for they had been well thumbed, perhaps by Harriot and his friends, perhaps by those of us who have come in search of him. In these papers, and in his research as a whole, I was intrigued that he generally focused only on the relevant facts. He was not given to wild speculations, either philosophical or mystical or religious – at least, not in the manuscripts that have survived. Historians have found a couple of additional documents about his reading matter, though, and they show that he read widely on all manner of scientific, re-ligious, and magical topics. This suggested to me that he was an open-minded person, albeit one of his time. In light of the focus on mathematics and physics in his manuscripts, however, it also hinted that he was in-stinctively aware that there were different ways of investigating and un-derstanding what he would have seen as God's creation. The most influential writings of Galileo and Kepler, for example, suggest a similar view: that the topics we now regard as scientific are best studied via experiment and mathematics, independently of philosophical traditions, intuitive notions, and scriptural prescriptions and prohibitions such as those that long lin-gered around the idea of an earth-centred universe and around infinitesimal quantities both mathematical and atomic. Newton made what turned out to be the definitive separation between these ways of knowing, although unlike Harriot he famously devoted far more manuscript pages to religious and mystical speculations than to mathematics and physics.

As for my feeling that Harriot was open-minded, this seems supported by the one work that he did publish: his remarkable 'first contact' record of life on Roanoke Island and surrounds. This area was in Ralegh's 'Virginia' but today's North Carolina, and Harriot spent a year there, in the famous First Colony. As is now well known, Harriot learned the local language, an Algonquian dialect, and his brief report shows – to me, at least – that he was unusually open to the people and their way of life. Some scholars, particularly those working in the New Historicist tradition of the 1980s and 1990s, have seen a different Harriot in this document (*A briefe and true report of the new found land of Virginia*): they have seen him as a canny propagandist for Ralegh's colonial and capitalist ambitions.[8] Indeed, propaganda was the

---

8 The negative view of Harriot expressed in the New Historicist period, most notably in Stephen Greenblatt's 1981 essay 'Invisible bullets', also accuses Harriot of 'appropriating'

pamphlet's motivation. But my own reading of the report aligns more with the view first enunciated by David Quinn and later amplified by others – in effect, that you can see Harriot's curiosity and humanity bursting through the spin.

In fact, brief as Harriot's report is, I felt he really comes alive here. For instance, he does not merely list the food and other local resources that Ralegh requested of him; he also describes the way the Algonquians grew, hunted, and cooked their food, and comments like a gourmet on such things as its taste and texture, and the companionable way the people ate together. He sounds interested, even engaged, as he describes their ceremonies and religious beliefs (which he does in a remarkably non-judgmental way in comparison with most contemporary Christian writers on indigenous religions). Then there are his admiring descriptions of the Algonquians' expertise in making fishing weirs, canoes, and pottery: Harriot comments that these latter hand-made items were as good as the English could make even with their more advanced tools. This kind of detail was surely not necessary if Harriot himself was interested only in the promotional document that Ralegh needed.

It is true nonetheless that Ralegh needed Harriot to show prospective colonists that the local people were not to be feared, so it is not surprising that some scholars have interpreted Harriot's report in a negative light. As I have indicated, we all bring our own subjectivities into it when we try to make sense of such long-ago material.[9] Nevertheless, Harriot's eye for ethnographic and geographical detail and his straightforward, non-judgmental presentation are generally widely praised.

So, too, are John White's drawings and paintings of Algonquian life. The relatively unbiased scholarship of both men is illustrated by comparison with Theodor de Bry's lavish compilation of Harriot's report and White's pictures, rendered as engravings by de Bry. I read a copy of the original

---

indigenous language and 'objectifying' the people. This ahistorical position has been countered well by later scholars, and Greenblatt himself, in a more recent work, has made a brief but much more positive judgement of Harriot's overall achievements. But not everyone who weighs in on Harriot has moved on from New Historicism. For instance, a recent reviewer praised the 1981 piece by Greenblatt (which accuses Harriot of wilfully deceiving and exploiting the Algonquian people) and went on to make further claims about Harriot's activities, concluding that his true legacy was not in mathematics, science, and ethnology but the service of magic and capitalism. This scholar provided no evidence for his claims, and I have seen nothing that supports them in Harriot's work or the recent scholarly literature.

9 We might have had a clearer picture of Harriot's own intentions, if most of his notes on America had not been thrown overboard by sailors during the First Colonists' dramatic departure. Harriot mentioned that he intended to write more fully about the Algonquian people themselves – a statement that corroborates, I believe, the assessment that he admired much about his hospitable hosts. But apparently he never did. As it is, he had to rely on his memory for much of his *Report*, which he wrote some two years after leaving America. The level of remembered detail suggests to me how deeply he must have engaged with the people, their land, and their language.

edition in the British Library. Entitled *America (Part I),* it was brought to the Reading Room from the King's Library, and I imagined for a moment that I might be holding the very copy that Ralegh perhaps presented to Queen Elizabeth I. Either way, it was a handsome, centuries-old book, with Harriot's original text supplemented by his captions to the engravings. But I was perturbed by some of de Bry's renderings, such as the 'Indian Man and Woman Eating; Their sitting at meate': de Bry Europeanizes and in the case of the woman sexualizes the poses and features, in contrast to White's original. I discovered that other scholars, too, have critiqued de Bry's presentation, which includes his own introduction, additional engravings, and a much more Eurocentric cultural framing than in White's and Harriot's original depictions.

And yet, while my own view is that regarding America, Harriot was far more a scientist and relatively enlightened human being than an ideological capitalist or colonialist, I do not mean to suggest that we should look away from the colonial tragedy in which he was complicit, even if unwittingly. As a non-indigenous person writing this section of Harriot's story, I knew I could not possibly do it justice: the disaster at Roanoke alone was heartbreaking, and the blame for it rested squarely upon the English. Harriot intimated as much in his report, and perhaps it is significant that he never returned to the New World – unlike his mathematical acquaintance Lawrence Keymis, who became one of Ralegh's trusted leaders in his later South American ventures. Instead, Harriot had taken up Northumberland's generous offer of patronage, which enabled him to spend his time more freely on scientific and mathematical pursuits.

As for how my view of these pursuits evolved during my journey with Harriot, I must admit that I find his achievements even more impressive than I had initially imagined. I do think Whiteside was on the right track when he made the assessment that started me on this path. And I do see Harriot as an early modern mathematician and 'scientist' of similar stature (although obviously not of the same influence) as his famous 'scientific' contemporaries.

I have already indicated that Harriot's critics balk at placing him in such august company as Galileo and Kepler. The debate seems to hinge on a particular definition of early modern science. Harriot's critics tend to categorize him as solely practical or 'applied', and they say that unlike Galileo, Kepler, René Descartes, or even William Gilbert, Harriot never found, and rarely attempted, a theory or 'causal' explanation. On the other hand, others, including myself, disagree with this characterization, whether applied to Harriot himself or as some sort of gold standard for 'great' early modern science and mathematics.

It has been a lively debate, as earlier volumes of Harriot Lectures have shown, and working through my own response to it was such an important part of the process of writing Harriot's biography that I shall outline my conclusions, and my bias, here. I must preface this by saying that when I embarked on this project it had not occurred to me that I might need to defend Harriot. It certainly had not occurred to me that attempting to re-store his reputation might be seen as some sort of mythmaking with na-tionalistic origins.[10] I simply thought that someone who made independent co-discoveries of landmark results was intrinsically interesting, because his work added new insight into the nature and history of scientific creativity. Anyway, since Harriot scholars have disagreed so strongly about his legacy, let me present my own sense of it here.

First, the bias: my interest in pre-modern science is forward-looking, in that I want to compare it with modern science; I am interested in seeds and influences on the development of science, as well as in differences between modern ways of thinking and earlier ones. Some of Harriot's critics, too, take this perspective when they suggest his contributions were less 'modern' than those of famous contemporaries such as Galileo.[11] As I have men-tioned, some historians object to this approach; they believe the focus should be on Harriot in his own time, not on the ways in which his dis-coveries relate to later ones. That is a valid perspective for some endeavours, and I certainly agree that it is important to recognize contemporary con-texts. But, like many others, I am particularly interested in Harriot because of the way he managed to work out, so early, so many results that have proved to be important in modern science and mathematics.

To get a sense of Harriot's intellectual standing in comparison with his most famous contemporaries, I shall take Newton's *Principia* as a turning point, the beginning of modern physics – indeed, of modern science. This is not to deny the historical complexities of such an assessment. It is simply to say that modern physics is a direct descendant of *Principia,* and that our conception of scientific explanation in general changed dramatically after its publication. For no one before Newton had ever created a successful, wide-ranging, predictive, and explanatory theory about a natural phenomenon.[12]

---

10 Perhaps the initial effort to search for Harriot's manuscripts was nationalistic, but I am uneasy with the charge of mythmaking expressed in the scholarly literature as recently as 2020; it criticizes unfairly the work of all the modern scholars who have sought – for various historical and scientific reasons having nothing to do with nationalism or mythologizing – to explore Harriot's life and work.

11 In my book I give more details on the debates, criticisms, and critics discussed in this essay, so I shall not reference them here. I also discuss in the book a couple of additional criticisms of Harriot. First, that unlike Galileo he did not challenge Aristotelian philosophy, which still held sway: aside from the issue of influence, I believe that Harriot's manuscripts show this view is unfounded. Second, that again unlike Galileo, he did not use his astronomical observations in the service of proving heliocentrism; in Harriot's Anglican Britain, however, heliocentrism was already widely accepted, thanks to Robert Recorde – who published a brief, pro-Copernican dialogue nearly 75 years before Galileo's famously 'heretical' *Dialogues* – and to Harriot's immediate predecessors John Dee, Thomas Digges, and Henry Savile.

12 Newton's theory of gravity, applied to the motion of planets and moons, is accurate to at least one part in 10 million. So it still does much of the heavy lifting in our understanding of motion within the solar system (including, of course, on Earth). Consequently, Newton's

Which is one reason I disagree with critics who 'demote' Harriot for not having a successful theory himself: by Newtonian standards, none of his peers did. But if *having a go* at understanding causes counts for something (and I agree that it does), then I also agree with those who argue we cannot, as his most vehement critics have done, dismiss Harriot's explanation of the formation of rainbows or his analysis of mechanical collisions – or his attempt at building an atomistic theory of light that would explain simultaneous reflection and refraction, something no one else before Newton had attempted to explain, as far as I am aware, and Newton did not get it right either.

I disagree with Harriot's detractors for another, perhaps more important reason, however. In their implicit assumption of a 'gold standard' of early modern science that excludes Harriot, his critics have tended to place the emphasis on explanatory and unifying paradigms; but this diminishes the significance of individual mathematical innovations and experimental results in the actual development of modern science. Newton's theory of planetary motion was initially contested because it did not explain the cause of gravity itself: if a causal hypothesis could not be embedded into a theory in a quantitative, testable way, Newton believed it was better to leave it out. (He never claimed action-at-distance as the mechanism of gravitational interaction.) So I do not think he would have been particularly impressed with the unproven speculations of Harriot's peers: the influences he singled out were experimental laws, expressed mathematically. He mentioned the mechanics of collisions (naming his older contemporaries Wren, Wallis, and Christiaan Huygens), Galileo's law of falling motion and his derivation of the parabolic path of projectiles, and Snell's law of refraction.[13] These were all results that Harriot discovered, too, not to mention his mathematical techniques and discoveries, some of which (notably the binomial theorem and algebraic interpolation) were rediscovered and further developed by Newton.[14]

There is another 'turning point' besides *Principia* that is relevant to assessing Harriot's standing in the pantheon of early modern scientific innovators: Descartes's treatises *La dioptrique, Les météores,* and *La géométrie.* They were

---

theory is subsumed into Einstein's, as a 'weak field approximation', rather than overturned by it.

13 Newton's influences: *Principia,* Book 1, the first scholium after the laws of mechanics and corollaries; and *Opticks.* He should have mentioned Kepler's laws of planetary motion, even though he ended up deriving them from his theory.

14 This suggests that had Harriot published, he *would* have influenced the direction of science. Some critics say there is no evidence that he realized the significance of his discoveries or saw them as part of an evolving body of knowledge. Yet his friend William Lower certainly did, urging him to publish his 'rarest inventions and speculations'. And, in addition to his dying wish that his friends publish the best of his 'mathematical papers', there is evidence that Harriot himself had begun to prepare several treatises for publication or distribution.

intended to illustrate the new philosophy of science laid out in his 1637 *Discours de la méthode* ... *[Discourse on the Method of Reasoning Well and Seeking Truth in the Sciences]*, to which they were appended; but some of the results in these appendices proved more enduring than the philosophy, which was ultimately eclipsed by Newton's method in *Principia*. These results included the first publication of the law of refraction, the first publication of a satisfactory quantitative explanation of the rainbow, and the foundations of 'analytic' (or algebraic) geometry in *La géométrie*, which is generally regarded as the earliest text expressed in essentially modern algebraic notation. But now let me compare these important results with Harriot's work on the same subjects.

It is well known that Harriot discovered the law of refraction before Descartes and even Snell; it is less well known that, unlike them (and unlike Ibn Sahl, too[15]), he generalized it to different pairs of refractive media *and* expressed it explicitly as a 'sine' law, not just in implicit geometric form. For he, too, was exploring a fledgling form of analytic geometry, and while Descartes and Snell were still doing trigonometry in the time-honoured geometric way, he seems to have been the first to intuit the notion of a trigonometric *function*. He showed this not only by expressing the sine law algebraically, but also, for example, in his explanation of the rainbow, which was both earlier and more sophisticated than Descartes's. Descartes relied on geometric ray tracing through raindrops to show that the light rays forming the rainbow cluster around (what we now call) the angle of minimum deviation as the ray passes through the droplet and out again to the viewer. But Harriot went further, finding a general algebraic equation for the angle of incidence corresponding to this minimum deviation. In addition, unlike Descartes Harriot discovered the phenomenon of dispersion (usually ascribed to Newton, who analysed it more fully); he even calculated the indices of refraction (as we now term them) for several different colours in different media – something no one else before Newton had done. It is work like this that marks Harriot as one of the best experimenters *and* mathematicians of his era. In fact, it is surely Harriot's posthumously published 1631 *Artis analyticae praxis* that is the earliest 'modern' symbolic algebra text.

Before I discuss Harriot's algebra, however, I should round off the previous paragraph by acknowledging that Descartes went much further than

---

15 Some popular commentators believe that the tenth-century Baghdad-based scholar Abu Sa'd al-'Ala' Ibn Sahl beat Harriot to the law of refraction, because of a recently found manuscript page showing the correct geometric diagram. In my book – and in my detailed popular article on Harriot's discovery of the law of refraction in *Cosmos*, no. 85 (2019) – I discuss and reference the discovery of this remarkable work and show why it does not qualify as the discovery of a law of physics. (Briefly, on currently available evidence, Ibn Sahl did not derive his result experimentally, did not generalize it, and did not express his hypothetical geometric ratio as a constant or find its numerical value.)

Harriot in other ways. For a start, he put together his ideas on optics and mathematics far more carefully, and actually published them. But he also made other original contributions, including – to take just one example – a geometrical explanation of the secondary rainbow, which Harriot's papers only hint at, as far as I can see. Then there is philosophy, a subject to which Harriot evidently made no contribution but in which Descartes excelled.

As for the *Praxis*, this was put together by several of Harriot's friends in accordance with his will, but it was not the finest exemplar of his mathematical ability, as examination of his manuscripts shows. Either way, though, here we see Harriot making advances in what we now call pure, abstract mathematics: he was far from just a 'mathematical practitioner' as some see him, applying mathematical techniques solely to solve practical problems. There was no practical imperative behind his methods for solving higher order polynomial equations, for instance. Even when his work does begin as a practical application, he often moves into deeper, more original territory.

For example, in the 1580s and 1590s both Harriot and Edward Wright independently constructed a complete table of 'meridional parts', needed for navigating along a rhumb line (or line of constant bearing). Over the next quarter century, Harriot developed new interests and new mathematics, some of which he ultimately used to construct an even more accurate table of meridional parts.[16] But to see this as the endpoint or primary purpose of Harriot's innovations is, I think, to underestimate the breadth and generality of his mathematical interests, influences, and creativity. It is Harriot, not Wright, who moves from mathematically constructing tables to finding a general equation for the lengths of a planar equiangular spiral and its three-dimensional version (the loxodrome or rhumb line). This was the first time anyone had found a formula for the length of any type of curve (other than a circle). And it is Harriot – not Wright, and not Galileo either, although he, too, is a key pre-calculus pioneer – who uses infinitesimal and other mathematical methods to derive not just the length of the spiral but also the same formula for the sum of meridional parts (which is equivalent to the integral of a secant) that is nowadays found using integral calculus. Harriot certainly did not have calculus, or even a fully general algorithm, and these new tables do suggest the epithet 'mathematical practitioner'. But I do not think anyone who has worked in mathematics, and who has seen Harriot's manuscripts, can doubt his passion for the subject *itself*. His discoveries, including his multifaceted derivation of this 'integral' sum, reveal a creative mathematician at work, someone who loved mathematics for its own sake and who was interested in fundamental principles, not just practical applications.

Indeed, when it came to navigation and the needs of his patron, Harriot's original tables, on a par with Wright's, would have sufficed. By 1614, when

---

16 In his 2000 Harriot Lecture, Jon Pepper reported that the accuracy of this second table of Harriot's was unsurpassed until the 1920s.

Harriot made his new tables, Wright's were already famous throughout Europe, and Ralegh, who had been in prison for a decade, was much less interested in exploration. Harriot, on the other hand, had made several other mathematical breakthroughs, which he applied in deriving the remarkable formula underlying his new tables. But here, too, I see Harriot the sophisticated mathematician, not just the 'practical' table maker, as he develops these results: the first proof of the conformality of stereographic projection, and the first generalization of the binomial theorem, which he takes beyond its traditional uses in solving equations and in practical combinatorics to the purely mathematical realm of non-integer powers – a feat of abstraction enabled by his unique use of fully symbolic algebra. He applied these binomial coefficients in a wholly new way when he pioneered the first general method of exact algebraic interpolation (rediscovered and refined into what is now known as Newton-Gregory forward-difference interpolation). Interpolation is useful in making tables, and Harriot no doubt used his new method in creating his 1614 table of meridional parts – but his handwritten treatise on the method focuses solely on the mathematical theory.

I could go on – but I hope I've conveyed here something of the breadth and depth of Harriot's researches, which place him among the greatest of the early modern pioneers of science and mathematics. I do not mean to imply that he was 'modern', out of his time; after all, not even the most famous of his contemporaries was. Like them, though, he made the kind of discoveries that would later lead to, or transform into, modern science and mathematics.

I do agree, however, that each of these geniuses had a different mind-set, a different emphasis, just as today's researchers bring different skills and emphases to their work. With Harriot and Galileo, for example, I prefer to see it as a case of apples and oranges, rather than a competition: each has different strengths and weaknesses, especially from the point of view of modern science. Each has similarities, too, which highlights their shared intellectual inheritance, as Matthias Schemmel pointed out in his 2004 Harriot Lecture. It is these differences and similarities that make comparing their work so interesting. In fact, although I have emerged from my time with Harriot even more impressed with his achievements than I initially imagined, my appreciation of the complex, often non-linear processes and layers in the history of science has deepened, too, thanks to my engagement both with Harriot and his contemporaries, and with the various debates and analyses of modern scholars.

For a biographer, it can be a difficult line to tread between personal engagement with one's subject and dispassionate presentation of the life actually lived. I spent years immersed in Harriot's life and work, trying to piece together all the disparate analyses and primary documents into an accessible whole, and trying to find a sense of *him*. With a man as inaccessible as Harriot, this was not an easy task. In my mind, though, he came alive, and I 'visited' him not just in the British Library (and Petworth), but also at Syon, where he spent his most productive years; I even paid my respects in Threadneedle Street, where he died and was buried. When the book went to press, it took me a long time to move on to other projects. I missed him!

Of course, no one can ever hope to present the definitive Harriot. He will always retain his mystery, and I imagine scholars will continue to search for him for a long time to come. No doubt they will discover new aspects of his life or work. They will certainly add more to our understanding of the culture in which he worked. In the meantime, I hope I have helped to bring him – and the hard work of so many Harriot scholars over the past seventy years – a little further into the wider world.

A Bibliography of Secondary Sources Relating to the Life and Work of Thomas Harriot Published since 2010

Polly Allingham

This bibliography is limited to material that has appeared since Daniel Jon Mitchell's bibliography of works published between 2000 and 2011 in Robert Fox (ed.), *Thomas Harriot and his world. Mathematics, exploration, and natural philosophy in early modern England* (Farnham, 2012), 243–47. Dr Mitchell's bibliography continued Katherine D. Watson's in Robert Fox (ed.), *Thomas Harriot. An Elizabethan man of science* (Aldershot, 2000), 298–303, which listed works published since John W. Shirley's *Thomas Harriot. Renaissance scientist* (Oxford, 1974).

Stephen Clucas maintains an up-to-date of list of publications relating to Thomas Harriot on the website of the Thomas Harriot Seminar, organized by Birkbeck Research Networks, Groups and Societies at Birkbeck, University of London: www.bbk.ac.uk/research/networks/thomas-harriot-seminar-group

Alghamdi, Mohammed Ghazi, 'Leaking letters: The case of Harriot's *report* and six letters of Englishmen', *Cogent Arts & Humanities*, 9, no. 1 (2022). 10.1080/23311983.2022.2093558.

Ares, J., J. Lara, D. Lizcano, and M. A. Martinez, 'Who discovered the binary system and arithmetic? Did Leibniz plagiarize Caramuel'?, *Science and engineering ethics*, 24 (2017), 173–88.

Arianrhod, Robyn, *Thomas Harriot. A life in science* (New York: Oxford University Press, 2019)

Arianrhod, Robyn, 'Refracted glory', *Comsos*, no. 85 (2019), 84–91.

Biggs, Norman, 'Thomas Harriot on continuous compounding', *BSHM bulletin. Journal of the British Society for the History of Mathematics*, 28, no. 2 (2013), 66–74

Biggs, Norman, 'More seventeenth-century networks', *BSHM bulletin. Journal of the British Society for the History of Mathematics*, 32, no. 1 (2017), 30–39

Biggs, Norman, 'Thomas Harriot on the coinage of England', *Archive for history of exact sciences*, 73 (2019), 361–383

Bloom, Terrie F., 'Borrowed perceptions: Harriot's maps of the Moon', *Journal for the history of astronomy*, 9 (2016), 117–122

Brioist, Pascal, 'Thomas Harriot and the mariner's culture: on board a transatlantic ship in 1585', in Fox (ed.), *Thomas Harriot and his world*, 183–200.

Bucciantini, Massimo, Michele Camerota, and Franco Giudice (eds.), *Il telescopio di Galileo. Una storia europea* (Turin: Guilio Einaudi Editore, 2012)

Chapman, Allan, 'A new perceived reality: Thomas Harriot's Moon maps', *Astronomy & geophysics*, 50, no. 1 (2009), 1.27–1.33

Clucas, Stephen, '"All the mistery of infinites": mathematics and the atomism of Thomas Harriot', in Rommevaux (ed.), *Mathématiques et connaissance du monde réel*, 113–154.

Clucas, Stephen, '"The curious ways to observe weight in water": Thomas Harriot and his experiments on specific gravity', *Early science and medicine*, 25 (2020), 302–327

Cormack, Lesley B., 'Handwork and brainwork: beyond the Zilsel thesis', in Cormack, *et al.* (eds), *Mathematical practitioners and the transformation of natural knowledge*, 11–35.

Cormack, Lesley B., 'The role of mathematical practitioners and mathematical practice in developing mathematics as the language of nature', *in* Geoffrey, Gorham, Benjamin Hill, Edward, Slowik, and C. Kenneth Waters (eds.), *The language of nature. Reassessing the mathematization of natural philosophy in the seventeenth century* (Minneapolis and London: University of Minnesota Press, 2016), 205– 228.

Cormack, Lesley B., Steven A. Walton, and John A. Schuster (eds.), *Mathematical practitioners and the transformation of natural knowledge in early modern Europe* (Cham: Springer, 2017)

Dawson, Scott, 'The vocabulary of Croatoan Algonquin', *Southern quarterly*, 51 (2014), 48–53

English, Neil (ed.), *Chronicling the golden age of astronomy. A history of visual observing from Harriot to Moore* (Cham: Springer, 2018)

English, Neil, 'Thomas Harriot, England's first telescopist', in English, *Chronicling the golden age of astronomy*, 1–10.

Falk, Dan, *The science of Shakespeare. A new look at the playwright's universe* (Fredericton, New Brunswick: Goose Lane Editions, 2014)

Fantazzi, Charles, 'Harriot's Latin', in Fox (ed.), *Thomas Harriot and his world*, 231–236.

Fox, Robert (ed.), *Thomas Harriot and his world. Mathematics, exploration, and natural philosophy in early modern England* (Farnham and Burlington, VT: Ashgate, 2012)

Fox, Robert, 'The many worlds of Thomas Harriot', in Fox (ed.), *Thomas Harriot and his world*, 1–10.

Fox, Robert, 'Thomas Harriot's Oxford', in Michela Malpangotto, Vincent Jullien, and Efthymios Nicolaïdis (eds.), *L'homme au risque de l'infini. Mélanges d'histoire et de philosophie des sciences offerts à Michel Blay* (Turnhout: Brepols, 2012), 103–109

Gatti, Hilary, 'Cosmological space between Copernicus and Newton', *Memoirs of the American Academy in Rome*, 58 (2013). 3–16.

Goulding, Robert, '*Chymicorum in morem:* refraction, matter theory, and secrecy in the Harriot-Kepler correspondence', in Fox (ed.), *Thomas Harriot and his world*, 27–51.

Goulding, Robert, 'Thomas Harriot's optics, between experiment and imagination: the case of Mr Bulkeley's Glass', *Archive for history of exact sciences*, 68 (2014), 137–178

Goulding, Robert, and Matthias Schemmel, 'The manuscripts of Thomas Harriot (1560-1621)', *BSHM bulletin. Journal of the British Society for the History of Mathematics*, 32, no. 1 (2017), 17–19

Groesen, Michiel van, *The representations of the overseas world in the de Bry Collection of Voyages (1590-1634)* [Brill's Paperback Collection] (Leiden and Boston: Brill, 2012). Abbreviated edition of the original work, published in 2008

Hacke, Daniela, 'Colonial sensescapes: Thomas Harriot and the production of knowledge', in Daniele Hacke and Paul Musselthwaite (eds.), *Empire of the senses. Sensory practices of colonialism in early America* (Leiden: Brill, 2018), 163–189

Henry, John, 'Why Thomas Harriot was *not* the English Galileo', in Fox (ed.), *Thomas Harriot and his world*, 113–137.

Hudgins, Carter C., 'Copper, chemistry, and colonization: the roles of non-ferrous metals at Jamestown (c.1607-10) and Roanoke (c.1585-90)', in Pope, with Lewis-Simpson (eds.), *Exploring Atlantic transitions*, 202–212.

Hyslop, S. J., 'The mathematics of collision and the collision of mathematics in the 17th century', Ph.D thesis (Indiana University, 2015)

Klingelhofer, Eric C., and James Lyttleton, 'Molana Abbey and its New World master', *Archaeology Ireland*, 24, no. 4 (2010), 32–35.

Klingelhofer, Eric C., and Nicholas M. Luccketti, 'Elizabethan activities at Roanoke', in Pope, with Lewis-Simpson (eds.), *Exploring Atlantic transitions*, 181–189.

Luccketti, Nicholas M., 'Copper carrieth ye price of all, or how Thomas Harriot may have saved Jamestown', in Eric C. Klingelhofer (ed.), *A glorious empire. Archaeology and the Tudor-Stuart Atlantic world. Essays in honor of Ivor Noël Hume* (Oxford: Oxbow Books, 2013), 1–11

Maclean, Ian, 'Harriot on combinations', in Fox (ed.), *Thomas Harriot and his world*, 65–87.

Molaro, Paolo, 'Thomas Harriot at the National Gallery?', *Astronomische Nachrichten*, 339 (2018), 103–108

Nicholls, Mark, 'Last act? 1618 and the shaping of Sir Walter Ralegh's reputation', in Fox (ed.), *Thomas Harriot and his world*, 165–182.

Pepper, Jon V., 'Thomas Harriot and the great mathematical tradition', in Fox (ed.), *Thomas Harriot and his world*, 11–26.

Pluymers, Keith (2011). 'Taming the wilderness in sixteenth- and seventeenth-century Ireland and Virginia'. *Environmental history*, 16, 610–63210.1093/envhis/emr056.

Pope, Peter E., with Shannon Lewis-Simpson (eds.), *Exploring Atlantic transitions. Archaeologies of transience and permanence in new found lands* (Woodbridge: Boydell, 2013)

Pumfrey, Stephen, 'Patronizing, publishing and perishing: Harriot's lost opportunities and his lost work "Arcticon"', in Fox (ed.), *Thomas Harriot and his world*, 139–163.

Reeves, Eileen, and Albert Van Helden, 'Turning the telescope to the sun: Thomas Harriot and Johannes and David Fabricius', in *Galileo Galilei & Christoph Scheiner. On sunspots*. Translated & Introduced by Eileen Reeves and Albert Van Helden (Chicago and London: University of Chicago Press, 2010), 25–34.

Rommevaux, Sabine (ed.), *Mathématiques et connaissance du monde réel avant Galilée* (Montreuil: Omniscience, 2010)

Schemmel, Matthias, 'Medieval representations of change and their early modern application', *Foundations of science*, 19 (2014), 11–34

Schemmel, Matthias, 'Thomas Harriot as an English Galileo: the force of shared knowledge in early modern mechanics', in Fox (ed.), *Thomas Harriot and his world*, 89–111.

Schuster, John A., 'Physico-mathematics and the search for causes in Descartes' optics—1619-1637', *Synthese*, 185, no. 3 (2012), 467–499.

Siegmund-Schultze, Reinhard, 'Johannes Lohne (1908-1993) revisited: documents for his life and work, half a century after his pioneering research on Harriot and Newton', *Archives internationales d'histoire des sciences*, 60 (2010), 569–596

Smith, Russell, 'Shining a light on Harriot and Galileo: on the mechanics of reflection and projectile motion', *History of science*, 53, no. 3 (2015), 296–319

Stedall, Jacqueline, 'Reconstructing Thomas Harriot's Treatise on equations', in Fox (ed.), *Thomas Harriot and his world*, 53–64.

Stedall, Jacqueline, 'Notes made by Thomas Harriot (1560-1621) on ships and shipbuilding', *Mariner's mirror*, 99, no. 3 (2013), 325–327

Stedall, Jacqueline, 'Thomas Harriot (1560-1621): history and historiography', in Wardaugh (ed.), *History of the history of mathematics*, 145–164.

Straube, Beverly A., '"A sure token of their being there": artefacts from England's colonial ventures at Roanoke and Jamestown', in Pope, with Lewis-Simpson (eds.), *Exploring Atlantic transitions*, 190–201.

Sullivan, Garrett A., Jr., 'Harriot, Thomas', in Garrett A. Sullivan, Jr. and Alan Stewart (eds.), *The encyclopedia of English Renaissance literature*, 3 vols. (Chichester: Wiley-Blackwell, 2012), vol. 2, 441–444

Swan, Diccon, 'The portrait of Thomas Harriot', in Fox (ed.), *Thomas Harriot and his world*, 239–241.

Tise, Larry E., and Michiel van Groesen (eds), *Theodore de Bry. America. The complete plates 1590-1602* (Cologne: Taschen, 2019)

Tise, Larry E., 'The 'Perfect' Harriot/de Bry: cautionary notes on identifying an authentic copy of the de Bry edition of Thomas Harriot's *A briefe and true report* (1590)', in Fox (ed.), *Thomas Harriot and his world*, 201–229

Vokhmyanin, Mikhail, Rainer Arlt, and Nadezhda Zolotova, 'Sunspot positions and areas from observations by Thomas Harriot', *Solar physics*, 295: 39 (2020). 10.1007/s11207-020-01604-4

Walton, Steven A., 'Technologies of pow(d)er: military mathematical practitioners' strategies and self-presentation', in in Cormack et al. (eds.), *Mathematical practitioners and the transformation of natural knowledge*, 87–113.

Wardhaugh, Benjamin (ed.), *The history of the history of mathematics. Case studies for the seventeenth, eighteenth and nineteenth centuries* (Bern: Peter Lang, 2012)

Williams, Travis D. (2016). 'Mathematical Enargeia: the rhetoric of early modern mathematical notation', Rhetorica, 34, 163–211.

Young, Sandra, 'Narrating colonial violence and representing new world difference: the possibilities of form in Thomas Harriot's "A briefe and true report"', *Safundi*, 11 (2010), 343–360

Index

Page numbers followed by "n" indicate notes.